CHAINSAWS
A HISTORY

CHAINSAWS
A HISTORY

David Lee
Produced in conjunction with
Mike Acres and the Chain Saw Collectors Corner

Harbour Publishing

Page 1:
Montreal-built Precision, 1951. *Collector Marshall Trover, photo Brian Morris*

Page 2-3:
1951 Disston DA-211, considered by some the best two-man chainsaw ever made. *Collector Mike Acres, photo Lionel Trudel*

This page:
A DeLorean of the chainsaw world with luxurious heated handles and a fluid clutch, the Vancouver-built Turbomatic 2-27 was an innovative 1950s one-off. *Collector Mike Acres, photo Lionel Trudel*

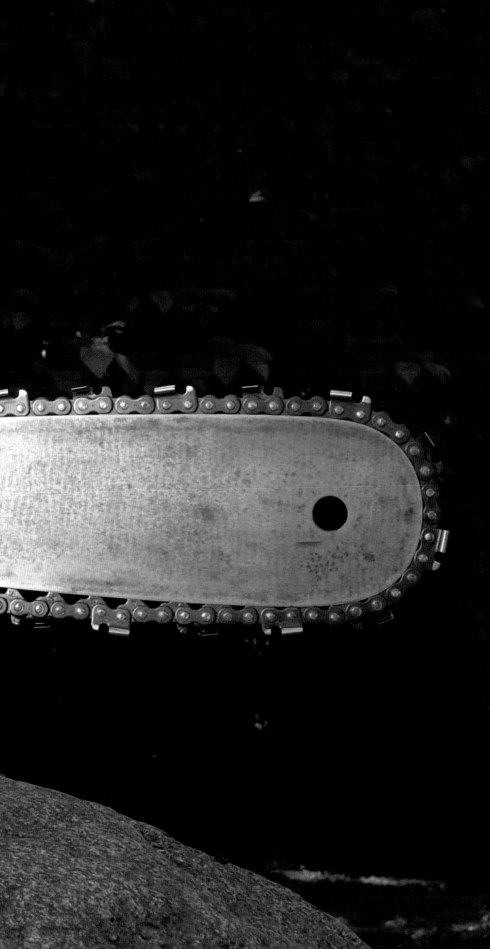

Table of Contents

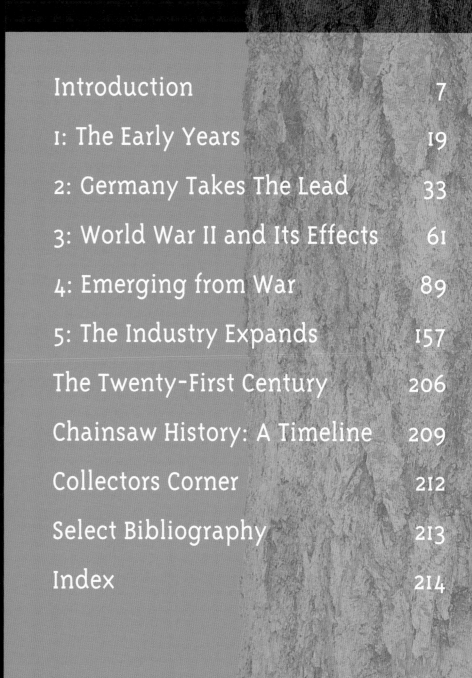

Introduction

I t rips and cuts, it splatters and drips, it raises the roof with its racket: the chainsaw is the most terrifying of all labour-saving devices. Its name has become synonymous with bad behaviour: chainsaw diplomacy or even chainsaw massacre (something it's only good for in the movies; in real life, as Mike Acres drily points out, "they'll hear you coming"). Yet for anyone whose life and livelihood has depended on felling trees, bucking logs or hand-sawing brush or timbers, the chainsaw inspires respect, loyalty, even affection. Your first chainsaw is guaranteed to drive you crazy—then you come to love it. This might be hard to understand for those who only see the chainsaw for its awful noise and smell, for the danger it poses to anyone who starts one up, and for its role in the modern crisis of deforestation.

A 1941 Titan Model A, a wartime knock-off of the German Stihl B2Z, perches on a Rayonier rail car. *Collector Marshall Trover, photo Brian Morris*

The 1952 Titan Sportsman had a fancy paint job designed to appeal to the non-professional.
Collector Marshall Trover, photo Brian Morris

Maybe some of those people will read this book and get their heads straightened around about this most misunderstood of power tools. Maybe not. There is so much chainsaw history to catch up on there is no space in this book for explaining just how wrong they are. This book is unapologetically addressed to that small but growing segment of the population who already know that the chainsaw is one of humanity's most admirable inventions and that the history of its development is one of technology's most unsung sagas.

I hope in these pages to document the ingenuity behind the development of the chainsaw as well as the many technical, economical and historical forces behind its growth and evolution. These chapters zero in on the genius of the many grassroots machinists and mechanics who have tinkered with an old concept (we have been trying to cut wood with a chain since at least the nineteenth century) and, through many different steps, false starts and dead ends, made it into the lightweight and efficient machine we know today.

Chainsaw buffs are attracted not just to the machine itself but to the special industry that created and developed it—an industry that has now virtually disappeared. Up until the late 1950s, the story of chainsaws was the story of how a few people in a workshop started a cottage industry that became a small business that became a big business (since the 1970s, the contemporary equivalent has been, if anything, the computer industry). However, it has been several decades since D.J. Smith set the foundations for Pioneer while working out of the basement of a Vancouver hotel, Claude Poulan hammered a truck fender into a bow saw, or Andreas Stihl schlepped across the Canadian Rockies and back to retrieve his last sample saw for BC loggers.

Above: Homelite's first chainsaw, the 1949 Model 20MCS, signalled the arrival of a major new brand. *Collector Marshall Trover, photo Brian Morris*

Left: This 1938 Stihl BDKH evolved into the primary German saw of WWII, the KS43. *Collector Marshall Trover, photo Brian Morris*

Those days are long gone. Over the years those one-man operations grew big enough to be acquired by huge multinationals—or in the case of Stihl, they have become huge multinationals themselves. The loss of independence and diversity in today's chainsaw industry is a recurring theme when one talks to chainsaw buffs. In North America there is also an underlying regret about how, in the process of growth and globalization, so much chainsaw control has gone overseas.

There were chainsaws everywhere when I was growing up in Mission, a small town in British Columbia's Fraser Valley, but I never paid much attention to them until I moved to Pender Harbour, a fishing, logging and resort community on BC's Sechelt Peninsula. Soon I bought a $40 used Pioneer saw—"sure, it works fine"—from a garage sale but even after the local chainsaw dealer fixed it, I couldn't start it. I tried and tried, pulled and pulled, choke out, choke in. Finally I drove the chainsaw to the local dump and threw it as far I could out into the largest pile of garbage—an act that appalls me now that I have come to treasure old saws as collectors' items, but at the time I was convinced I hated the machine and all its noisy, temperamental kind.

As if to punish me, fate soon found me working for the local equipment rental business, which sold and serviced chainsaws. I started out thinking there would be enough to do in the shop that I could manage to avoid the accursed machines. But it was not to be. One

McCulloch's revolutionary 1949 3-25 was the first one-man saw that would run on its side, thanks to its diaphragm carburetor. It helped establish McCulloch as the leading chainsaw maker in the 1950s. *Collector Marshall Trover, photo Brian Morris*

The English Aspin P-54 was J.H. Sankey & Son's 1953 entry in the one-man saw stakes. *Collector Marshall Trover, photo Brian Morris*

The 1948 Bluestreak was Seattle-based Titan's most successful two-man saw. *Collector Marshall Trover, photo Brian Morris*

With its sleek design and reliability, the one-man Super Pioneer was another winner for Vancouver-based IEL in 1951.

Collector Mike Acres, photo Lionel Trudel

Opposite page and right: The David Bradley 50 was built for Sears Roebuck by Reed-Prentice using engine parts made by McCulloch. *Collector Marshall Trover, photo Brian Morris*

Below: The 1952 Torpedo was Vancouver-based PM's entry in the one- or two-man market, but it didn't find its target. *Collector Marshall Trover, photo Brian Morris*

day my boss, as usual juggling several customers at once, paused in one of his anecdotes, turned to me, placed a beat-up McCulloch in my hands and said, "Take this out back, take off the air filter"—he pointed to a couple of screws—"and let's see how it looks."

That was the beginning of my education as a chainsaw mechanic, and I still get a thrill of pride when I remember the first time I was able to not only fix a saw on my own, but advise the highly sceptical customer on how to stop his problem from recurring (in this case, it was to blend the proper ratio of two-stroke mix oil to gasoline in a separate container before fuelling the saw—not to half-fill the tank with straight gas, then top it up with chain oil). Soon I was honing my skills on Stihl, Homelite, Husqvarna, Craftsman, McCulloch and a whole bunch of brands I'd never heard of. I even managed to get the occasional old Pioneer going, soothing my guilt over the beast I'd pitched into the dump—a beast I probably could have brought back to life had I known then what I know now.

I had hated chainsaws. Now I loved them.

By 2004 I had moved to Hamilton, Ontario to do post-graduate work at McMaster University. One day in a massive east-end junkshop I came across an old chainsaw I'd never heard of before. What the heck, I asked myself, is a *David Bradley*? The saw brought back nostalgia for the years in Pender Harbour and it piqued my curiosity. I thought about that

Norway's Jo-Bu Junior looked a bit of a throwback for 1952, but it was just right for Scandinavian forest work and sold 40,000 units. *Collector Mike Acres, photo Lionel Trudel*

old chainsaw, searched around on the internet and started wishing that there was a book I could go to, a big, illustrated history of the chainsaw. I found a number of how-to books on chainsaw use, maintenance and safety, many of them excellent, but to my amazement there was nothing like the kind of comprehensive history I was looking for. It occurred to me that as a writer who knew a thing or two about chainsaws, I was in a position to do something about this unaccountable void in the literature.

Of course, writing a true comprehensive history of the chainsaw would be like writing a true comprehensive history of the automobile—it would have to be about twenty thousand pages long, so the writer's question "What to put in?" pales beside the larger question of what to leave out. There are literally hundreds of manufacturers past and present. Information has been often hard to come by. In the past, as now, chainsaws tend to be a subject not brought up in polite society and many of the machine's milestones have been sparsely documented. What information does remain is not, by and large, to be had from manufacturers, who are in business to sell this year's models and who with few exceptions do not see themselves as the caretakers of an often-fascinating industrial history.

Some sawmakers get a lot of space in this book and some don't get much. This is not always a reflection of the manufacturer's prominence but simply of how much information I have been able to find. Maybe this book will encourage some closet chainsaw enthusiasts to come forward with what they know and some of the industry's veterans to write down their experiences. If anything, I'm hoping this book provides an overview of the chainsaw and its proliferation throughout the last century, and an appreciation of the creativity and enterprise that have produced a host of different and ever-changing designs over the years.

The project may have come to nothing had I not had the good fortune to hook up with Mike Acres, a long-time chainsaw buff and collector who runs his Chain Saw Collectors Corner website out of Burnaby, BC. I have depended heavily on the contributions of Mike, as well as his friend and colleague, the super-collector Marshall Trover of Renton, Washington. Dave Challenger, Wayne Sutton, the late Peter Knight, Nico Henkens and others literally from around the world have all provided invaluable assistance in detailing the history of this powerful, cantankerous and often misunderstood machine. My grateful thanks to all.

David Lee
Hamilton, Ontario
August 2006

In 1925 this 1-hp Wolf electric was a true technological wonder.
Collector Marshall Trover, photo Brian Morris

The Early Years

Experiments in Power Saws

"In one place the woodmen had been at work on Saturday; trees, felled and freshly trimmed, lay in a clearing, with heaps of sawdust, by the sawing machine and its engine."

—H.G. Wells, *The War of the Worlds*, 1897

W ell into the twentieth century, the prevailing tool for sawing and bucking in the woods continued to be the handsaw: either the "misery whip"—the two-man cross cut saw—or various one-man Swede, buck and bow saws. These tools were inexpensive and lightweight but murderously labour-intensive. Paired with the axe, wedge and springboard, the cross cut saw could eventually bring down any tree that was ever grown.

A steam-powered drag saw. *Vancouver Public Library, VPL 4995*

English inventor A. Ransome's portable steam-powered saw was quite a success in 1860.

Some inventions, like this Rube Goldberg affair, made the simple handsaw look good.

So, was H.G. Wells' turn-of-the-century "sawing machine" as much a product of his imagination as invading Martians?

Surprisingly, it wasn't. Machines for cutting wood were being designed, built and even sold commercially decades before the modern chainsaw came on the scene. There was a huge incentive to invent them. Even today, anything to do with cutting and sawing wood is hard work, and before mechanization it was gruelling. Canny woodcutters were always on the lookout for some better way of doing things.

A design from 1856 resembled the modern pipe cutter. It consisted of a cogwheel big enough to fit around a tree. On the inside of the cogwheel was a strong blade, and on the outside was a hand crank. As the crank was turned the blade circled the tree trunk cutting deeper and deeper into it.

A few years later came a more successful design, again based on the hand-driven rotary principle. With crank handles, two men turned a large flywheel to power a saw in a back-and-forth motion.

Around this time, in 1858, the first saw chain was patented in the USA. But as yet there was no way to move the chain at high speed for efficient cutting. The goal of most innovators and inventors was still to duplicate, and amplify, the known and dependable back-and-forth (reciprocating) motion of a simple handsaw.

None of these inventions became successful commercial ventures until the English inventor A. Ransome made a steam-powered saw in 1860. From a wood-fired boiler set up on-site, pressure hoses led to several heavy portable saws. Each saw had a single-cylinder motor, its piston connected directly to a reciprocating saw blade. The Ransome's disadvantage was the need to transport a heavy boiler into the woods (illustrations show it horse-drawn, on wheels) and keep it supplied with water and fuel. Its advantage was that freed from their power source, the saws could be designed to fit the needs of working in the woods. With a spindle and cog wheel, the blades could be adjusted horizontally or at an angle for falling, and vertically for bucking.

This saw was quite a success—probably H.G. Wells' "woodmen" were using a Ransome—

and at the turn of the century it was being used in Europe and Africa. But it was still a heavy piece of equipment that in rough country was unmanageable. Even without the boiler or the bucking accessories, each saw on its own weighed 600 pounds (273 kg). No wonder the English woodcutters of Wells' novel had no qualms about leaving their gear standing out overnight! When gas-powered saws built along these lines enjoyed brief popularity in the twentieth century, this sort of mechanized reciprocating design came to be known as a "drag saw."

The First Attempts with Chain

The end of the nineteenth century also brought the practical beginnings of the chainsaw. This is a time when the new technologies of electricity and the internal combustion engine were bringing a tumult of invention. Small-scale, portable power was being applied to every manual task from washing to cycling to milking cows. It was inevitable that wood-cutting would be included.

As motors improved and proliferated, the concept of the one-directional cutting chain came into its own. An Oregon inventor patented a design for a "sawing chain and frame" in 1897, but it never went into production. Evidently in 1905 the Ashland Iron Works in Oregon offered a pneumatic chainsaw that "met with limited use," but did not go anywhere commercially.

However, a gasoline-powered chainsaw was demonstrated at Eureka, California in 1905. "The name of the inventor is lost but two men bearing the names of Benz and Hendricks are associated with the event." Driven by a two-cylinder, water-cooled engine, the machine drew its fuel and water supply from tanks that were nailed to the tree trunk above it and "removed when the tree was about to go over." This must have made for some tense moments! The machine had no bar; its chain circled the trunk and cut inwards from every side. Understandably, the saw was never commercially produced, though it evidently was able to cut through a log ten feet in diameter in four and a half minutes.

A few years later, the cover of *Scientific American* featured a chainsaw with an enormous bar, at least eight feet long, bucking its way through a fallen forest giant. The caption read simply "Cutting a redwood tree with a saw driven by an engine." A short article titled "A Power-Driven Saw," buried on page eighty-six of the issue gave somewhat more detail, though the term "chainsaw" was never used. Rather, it was a "continuously running flexible saw," arrived at by attempting to build a bandsaw that would cut down trees without the encumbrances caused by the fact that "the band necessarily runs in two planes."

A Californian, R.L. Muir had developed this "endless cross-cut saw." Although the bar and chain assembly looked remarkably similar to those of the chainsaws that would start to appear in the 1920s, there the resemblance ended. Intended for bucking the huge trees

The whip saw has been producing sawn lumber since biblical times. *University of Washington Special Libraries, Special Collections HEG037, photo Eric E. Hegg*

R.L. Muir's "endless crosscut saw," one of the earliest gas-driven chainsaws, early 1900s. *Collector Marshall Trover*

The Sector chain.

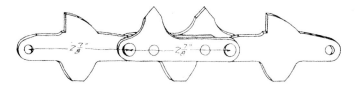

The Wolf "Brute."

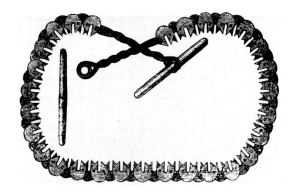

Early sawing chains: hand-powered.

Below: Like most early chain, the Wolf chain was modelled on the cross cut saw.

of the redwood forest, the saw was designed to be run by a gas engine or an electric motor mounted on a skid alongside the tree trunk.

It seems likely that this device was tried in the woods and abandoned. Supposedly, the larger model of the "endless cross cut saw" could saw through a log five to seven feet thick in less than ten minutes. Still, mounting the crane-like bar assembly and the motor onto a scaffolding sufficient to bear their weight, and having to anticipate the extra stresses generated by the running engine, must have taken considerable time. One can picture a couple of indignant Swedes running ahead of the apparatus and getting all the bucking done while Muir and his helpers were still hammering together the skid. Although the claims in 1910 were that Muir's device had enjoyed "remarkable success in the vast redwood belts of Mendocino and Humboldt Counties in California," and that it could be run horizontally, vertically or on an incline, it more or less vanished into history.

Why Mess With Chain?

Made up as it is of hundreds of small moving parts, a saw chain is a complicated thing to produce. Onto this unstable base, effective cutting teeth must somehow be attached. A chain must be held rigid by being stretched around a stiff blade known as a guide bar or just plain bar. This bar must have a channel for the chain to run in and it must be continuously lubricated to reduce friction and wear. There must be a sprocket system for driving the chain at optimum cutting speed. It took saw designers decades of experimentation before they were able to perfect the cutting chain and bar. Why did they bother?

In terms of power mechanics, it had long been known that continuous rotation in the same direction was more efficient for cutting than the reciprocating back-and-forth motion of the handsaw or drag saw. This principle had been learned in early sawmills that had begun using water wheels to power reciprocating saws as far back as the middle ages. Because a reciprocating blade must constantly stop and reverse direction, it can never build up the cutter velocity a rotating blade can. Constantly stopping and reversing also wastes energy and creates vibration, which makes the reciprocating saw harder to guide for accurate cutting.

Inventors brought this knowledge with them when they turned to the portable power saw, but it wasn't practical to adapt the circular blade for tree falling. Since a circular saw can only cut half as deep as the saw blade is broad, sawing down large trees meant carrying blades of unwieldy size into the forest, with heavy engines to turn them—though some tried it.

These difficulties forced early portable saw manufacturers to turn to slow, inefficient reciprocating saws, but visionaries kept dreaming of a device that would marry the smooth cutting speed of a circular blade to the practical linear shape of the reciprocating saw. They continued to come back to the concept of a linear blade with a line of teeth moving around its outer edges—something that could only be achieved with a looped chain running in a track. Although the mechanics were difficult, the concept proved sound. From its earliest working models, the chainsaw cut smoother and faster than the reciprocating saw.

World War I: Saws Needed

War accelerates technological change, and the shortage of manpower, and increased demand for wood caused by World War I brought new ideas about mechanizing wood-cutting. An attempt was made to perfect a "wire rope tree feller," using a steel cable to cut through a tree by friction rather than by sawing per se. Slightly more advanced was the spiral, fluted design of the steel "rope saw," intended to cut in both directions. Attached by cables to a double drum winch, the rope saw cut as the wire rope was run back and forth between the two drums of the winch.

Experiments were made using electricity to heat a wire element to burn through wood. Another device, the "power feller," was a long roller equipped with cutting teeth and mounted on the end of a shaft. Swinging in a horizontal arc, it would eat its way through a tree. Another design didn't cut at all, but used a series of augers to bore into a tree below ground level, felling it stump and all, but it never progressed beyond the planning stages. The trouble, expense, danger and effectiveness of these devices never managed to match the prevailing efficiency and economy of teams of men with misery whips.

After the basic reciprocating design inspired by hand saws, circular saws seemed by many to be the next logical step. In Russia, an electric circular saw was invented for felling trees. In France, a similar saw was mounted on the front of a hand-drawn carriage that had to be rolled from tree to tree.

The Hamilton saw, 1861.

The Multnomah steam saw in action. The old-timers tried everything to avoid picking up that handsaw, but in most cases their inventions caused more work than they saved. *Photo by Darius Kinsey, D. Kinsey Collection #18850, Whatcom Museum of History & Art, Bellingham, WA*

23

The Holt stump saw: circular saws cut like crazy, but they were cumbersome and dangerous in the forest.

Yet the reciprocating model was still the strongest contender. Compared to the problems posed in cutting with a chain, the blade was a tried and trusted tool. Also, many woodworkers were already skilled in the art of sharpening saw blades (to this day, the ability to properly sharpen chainsaw cutters is a rare and valued commodity) and to many, the best bet was the mechanized, solid-bladed "drag saw," the gas-powered descendant of Ransome's steam saw. This heavy, motorized version of a cross cut saw, mounted on a cumbersome frame, was clunky but dependable. It certainly made a significant contribution to the fact that today the legendary California redwoods are few and far between.

But the redwoods were huge trees on a landscape that was by and large gently sloping, relatively dry and free of underbrush. Big and heavy as these saws were, it was still possible to get them to the trees, and the time spent setting up a saw was well invested in the amount of wood recouped when such a large tree was harvested.

Drag Saws

The contender to chainsaws in the early days were drag saws—essentially large, gas-driven reciprocating saws. As pioneering logging executive Jack Challenger described it, "The chain saw can be built lighter, more compact, more portable, and with a much higher cutting speed than the drag saw. The drag saw, on the other hand, is more rugged and simple than the chain saw, consequently more dependable in operation."

Beginning in the late 1920s, drag saws enjoyed considerable popularity and continued to be marketed into the 1950s. Leading manufacturers were Stover, Wade, Witte, Ottawa, Vaughan, Canadian and Eureka, offering models equipped with blades from five to sixteen feet long. The one-piece blade, similar to a cross cut saw, was cheaper to buy, simpler to maintain and more durable than the chains of the time; however the saws themselves, large, heavy and needing to be carefully mounted to fell a tree, were classified among the power tools diplomatically called "semi-portable." Their best and most lasting application was for bucking firewood. It took a hefty lift to get one positioned over the woodlog and dogged in place, but once the cut was started the operator was free to prepare the next cut, split wood, or otherwise multi-task. A 1937 ad for a 4-hp, 265-lb. Pacific promises it will saw through a four-foot log in five minutes and produce 25 cords of firewood per day.

Around 1937 Eureka's designer, Joe Pesola, introduced the Pesola drag saw that "could be reduced into two parts in less than a minute's time for moving to a new location," powered by a 2-hp Briggs & Stratton four-cycle engine. A promotional photo shows a man of no great stature casually hanging onto the "power plant in one hand, saw frame in the other." However since the Pesola weighed 130 pounds, the guy in the photo must have been working hard to appear under no great strain.

Steam donkeys needed a lot of firewood and solid-bladed drag saws were just the thing to cut it. *University of Washington Special Libraries, Special Collections CKK01604, photo Clark Kinsey*

One of the first genuine chainsaws made for forest work seems to have been the Sector, invented by A.V. Westfelt in Sweden in 1919. As with the Ransome invention, the Sector was only viable because the saw itself was separated from its power source. Positioned nearby, a two-cycle outboard motor engine that had been adapted for the purpose ran the saw via a flexible driveshaft. Although it had some degree of mobility, the Sector was hampered by its dependence on the separate engine that had to be constantly repositioned as the operator worked on the tree being cut.

Just after the war, knowing nothing of the Sector, a farmer in Dauphin, Manitoba, named James Shand was using a horse team to draw barbed wire around his quarter-section of land. When he was done, he found that the horse-drawn wire had sawed through a seven-inch oak fencepost.

Shand was a trained millwright and in his home workshop, he extended the barbed-wire idea, fitting cutting teeth into his son's bicycle chain. His patent drawings show an impressive design, his bar foreshadowing the modern chain guide bar, but once again the power source was a problem. Using a flexible Bowden cable, Shand attached his new saw to a one-cylinder gasoline engine. He used one on the job at the Manitoba Bridge and Iron Works, and in 1919 took two working models to British Columbia, hoping to find interested manufacturers, but the Shand saw's 24-inch (60-cm) bar, fine for milled lumber and prairie trees, was too small for west coast logging. The heavy gas engine also limited its portability.

The internal combustion engine was still a young technology, one that lacked the portability and power that a mechanized saw would need. This was not true of electric motors, which had superior versatility and power at this stage.

The portable gas-powered drag saw was an effective bucking device once it got set up. *University of Washington Special Libraries, Special Collections, CKK0897, photo Clark Kinsey*

The Sector was connected by a flexible shaft to a converted outboard motor. *Collector Marshall Trover*

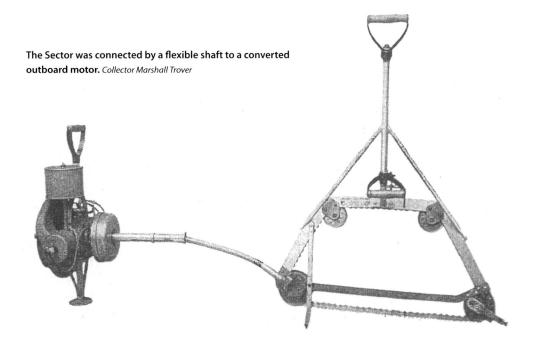

The Sector, developed in Sweden in 1919, was the first fully portable gas-powered chainsaw. *Collector Marshall Trover*

In 1920 Charles Wolf launched the first viable chainsaw company with a series of electric and pneumatic models.
Collector Marshall Trover, photo Brian Morris

WOLF

Charles Wolf and the First Successful Chainsaw

The first viable commercial chainsaw firm in North America was founded in 1920 by the engineer Charles Wolf. It had been over sixty years since the first United States chainsaw patent had been issued in 1858, but until Wolf appeared on the scene, no viable chainsaw had been developed. As the inventor's son, Jerome Wolf, wrote many years later:

An examination of patent and other records between 1858 and 1920 indicates that most people interested in chain saw development were not qualified. During this period, no doubt there were those who were qualified, but they apparently were not interested. My father was both interested and qualified. His mechanical and electrical engineering experience was unusually comprehensive and varied.

Born in 1871, Charles Wolf was an active innovator from an early age. In the 1890s, he worked with John Holland in developing a modern submarine prototype for the US Navy. Later, he was involved in the consolidation and modification of Henry Huntington's pioneering Los Angeles electric street railway and interurban systems. "Subsequently," Jerome wrote, "he built other electric railways, dams, bridges, tunnels, lights and water systems and whole town-sites."

By 1910, Charles Wolf was chief engineer for F.A. Blackwell, who owned the Panhandle Lumber Company at Spirit Lake and the Blackwell Lumber Company at Lake Coeur d'Alene, Idaho, as well as an interurban electric line between Spokane and Coeur d'Alene. Wolf was in charge of Blackwell's newest Panhandle project at Ione, Washington: the world's first electrically operated sawmill. He hired an ambitious electrical engineer at the Washington Water Power Company, Frank Redman, to work with him on designing and building the project.

By the time the mill opened for business in 1911, Wolf was already dreaming up another technical innovation. In 1906, several mechanics and machinists at an Idaho sawmill, the Potlach Lumber Company had devised a crude log deck chainsaw. This machine was used only on the premises and was never patented or mass-produced. However, since seeing it in 1908, Charles Wolf had never been able to get it out of his mind.

This 1-hp electric was Wolf's second offering.
Collector Marshall Trover, photo Brian Morris

The Wolf chain featured clusters of cutting teeth separated by rakers, similar to a cross cut saw. *Collector Marshall Trover, photo Brian Morris*

The Coeur d'Alene sawmill had a recurrent problem handling the 40-foot logs that were stored or boomed in the lake. Other sawmills used circular saws to trim logs, but it took a huge circular saw to cut through the biggest butts. Wolf thought the problem could be solved by a machine similar to the chainsaw he had seen at Potlach.

Wolf and Redman designed a new chain and a new machine for sawing logs to length at the Blackwell Lumber Company. The cutting chain, the sprocket that gripped and propelled it and the bar that supported it all had to be custom made at Peninsula Iron Works in Portland.

Wolf didn't bother to patent the new saw, or the chain that he and Redman had designed for it, because he was already thinking of ways to take these developments a step further. By 1920 he had patented and the Peninsula Iron Works was manufacturing the world's first commercially successful portable chainsaw, the Wolf Electric Drive Link Saw.

The saw came in three models. Model A had a 24-inch (60-cm) bar and weighed 70 pounds (31.7 kg), Model B had a 36-inch (91-cm) bar and weighed 80 pounds (36 kg) and Model C had a 48-inch (122-cm) bar and weighed 90 pounds (40.8 kg). All models were powered by a 1.5-hp electric motor that could either draw power in-house in a modern electrified sawmill or in the field from a portable generator.

Wolf had also modified the cutting chain, but its teeth were still set up in the traditional cross cut saw configuration with clusters of pointed cutting teeth separated by hooked rakers. Like the cross cut saw it was copied from, it cut in either direction, so that when the cutters and rakers in one direction began to dull, the chain could be taken off and reversed to prolong the cutting session. Such a chain would become the industry standard until the 1950s.

The first models were successful, but to facilitate production, Wolf looked to the east. He licensed his new Wolf saw and chain to the Reed-Prentice Corporation of Massachusetts, a well-established US manufacturing concern. Taking advantage of the new facilities, the electric Wolf models were augmented by an air-powered saw in 1927.

Strangely enough, the first buyers of the Wolf saw weren't in the forest industry. Indeed, Wolf didn't particularly aim these new products at the logging market. Instead, he targeted the construction trade. Because his chain made such a straight and smooth cut, his saws could be used in framing and finer work with milled lumber. Builders found that they could use a Wolf saw to frame single large timbers or to cut flooring, stair stringers, or rafters. His first customers included private contractors, the military, shipyards, mines and public utilities as well as sawmills. The pneumatic model could be operated by divers for cutting pilings and other timber underwater.

Meanwhile, the logging industry was not exactly beating down Wolf's door. There were few jobs anywhere harder than sawing with the "misery whip," but fallers and buckers

Wolf's chain was known for its precise cutting.

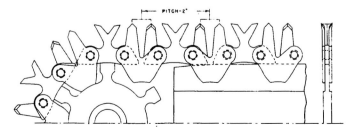

Wolf chainsaws were effective and versatile cutters, but their reliance on generators or compressors limited their portability.

Collector Marshall Trover, photo Brian Morris

(Picture of U. S. Frigate Constitution and Wolf electric chain saw reprinted from *Engineering News-Record*, January 3, 1929.)

Wolf Electric Machine—Introduced 1920

WOLF Portable Timber Sawing Machine

Collector Marshall Trover

were used to it, and brought a lot of professional pride to their craft. They also felt some trepidation about the new skills these new machines would demand. In fact, they feared that the adoption of the chainsaw would introduce a whole new approach to working in the woods—one that could soon make them obsolete.

The new machines were also expensive—an expense compounded by the portable generator that was needed to run them in the woods. Nevertheless, the sheer cutting power promised by the Wolf electric chainsaw attracted attention. In 1929, the Maine Logging Company of Belfair, Washington tried one—and immediately found that the trouble and expense was more than offset by their savings in falling and bucking costs. A year later, Weyerhaeuser Camp Two at Klamath Falls and the Booth-Kelly logging operation near Eugene had similar results. Soon Wolf was enjoying a brisk business—mostly to government agencies and the larger firms who could afford these big, expensive machines. This was a limited market, but a lucrative one. Wolf employed few salesmen and, as Jerome Wolf reminisced, "a collection loss was rare."

Selling was very different in those days. Once the Wolf enterprise really got off the ground, they employed a well-printed, generously illustrated booklet that doubled as a catalogue and as an operation and repair manual. As Jerome Wolf later pointed out:

> *Today there are thousands of distributor-dealer outlets specifically for chainsaws and related items. In the Wolf era, there were only a handful of industrial supply firms who could be relied on. Others stocked no parts and were otherwise unsatisfactory. As a result, the majority of our customers were entirely dependent upon the manual for service. Numbered photographs showed the various items so clearly that there was little chance of error in identifying parts. By means of the manual, customers all over the world were able to service their machines by mail with little or no difficulty.*

Wolf had few service headaches until the 1930s, when he ventured into the still-new territory of gas engines. Meanwhile, he found that aided by the detailed manual, most customers could easily service their own pneumatic or electric machines. "The AC electric motors," Jerome maintained, "operated for years without requiring repair or replacement parts."

Despite the success of the electric and pneumatic models, the potential of a self-contained gasoline-powered saw was obvious. In 1931, the Reed-Prentice plant produced its first Wolf saw powered by a four-cycle, 4-hp, air-cooled gasoline engine. Like the other first gas-powered saws, this was a big step forward, but heavy at 80 pounds (36 kg). Due to its weight, and the limited RPMs of the four-stroke engine, this was another saw that never got beyond the prototype stage.

By 1933, Wolf was selling an electric saw with a 16-inch (40-cm) bar for $495. The largest electric model, with a 48-inch (122-cm) bar, cost $860. Pneumatic and gas saws were in the range of $595 to $975. The most popular Wolf saws, the electric and pneumatic with a 24-inch (60-cm) bar, retailed for $560 and $645 respectively.

By 1936, however, the German companies Stihl and Dolmar were both producing high-quality gas saws that were competitive with Wolf in weight and price. When the first of these arrived in North America in the mid-1930s, the writing was on the wall for the Wolf saw.

Wolf's patents on his saw chain and his original machine expired in 1942, and so did his licensing agreement with Reed-Prentice. For the time being, Wolf was unable to produce saws, but he made a deal with a Portland, Oregon firm to produce chain and sprockets. At the very least as a parts source, his business would still have been viable—if the United States hadn't entered World War II. According to Jerome Wolf, "We began to develop a nice business supplying the Wolf machines already in the field and with others. However, the Portland manufacturing firm's war business increased to the point where we could no longer be accommodated."

Forced out of the Portland facility, the Wolfs waited out the war, but the next thing they knew, it seemed that everyone was making chainsaws. By 1945, there were so many new manufacturers that it became clear that the company's day had come and gone.

The 1942 Wolf pneumatic was produced for the US Army by Reed-Prentice. *Collector Marshall Trover, photo Brian Morris*

This Wolf pneumatic got a lot of use during WWII. *Collector Marshall Trover, photo Brian Morris*

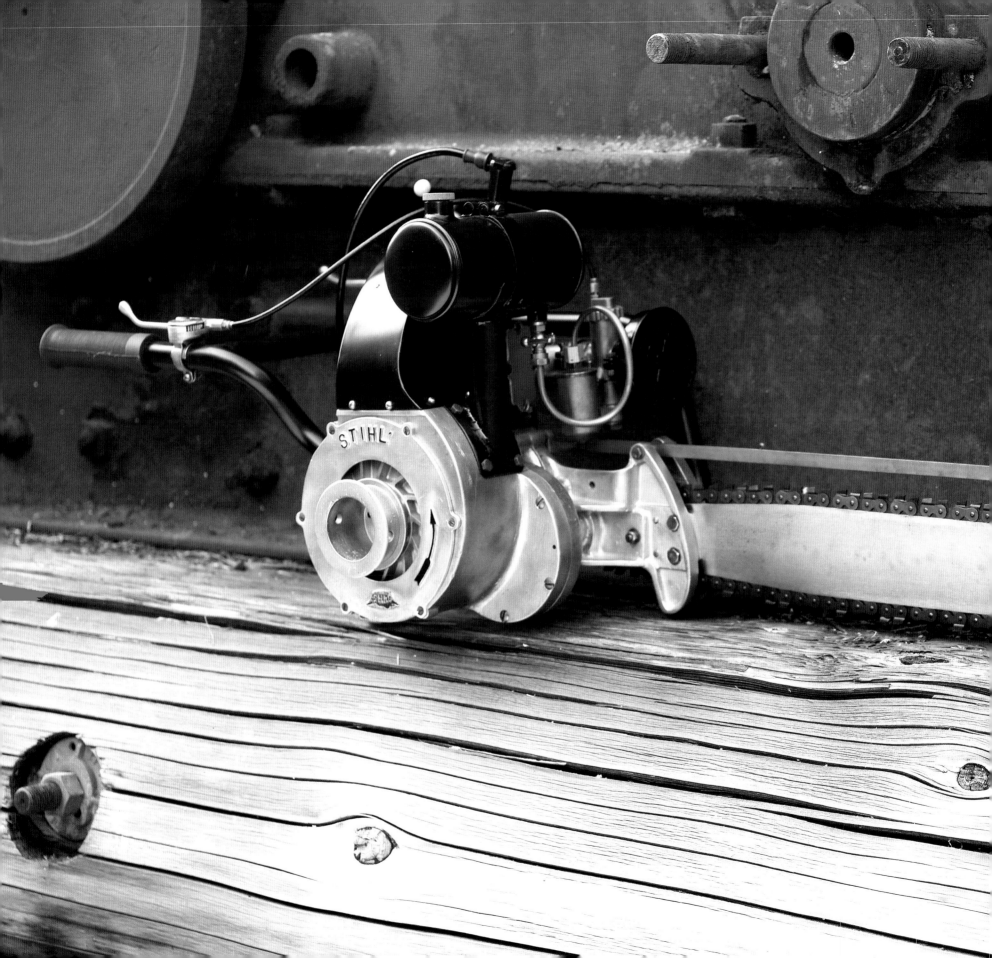

German inventors Andreas Stihl and Emil Lerp launched the chainsaw revolution with the first practical gas-powered saws in the mid-1920s. This is a developmental version of the 1930 Stihl Model BK. *Collector Marshall Trover, photo Brian Morris*

2

Germany Takes the Lead

Going for Gas

"In the twenties and thirties most of [Wolf's] competition came from Europe. There were Rinco, Erco, Dolmar, and Rapid from Germany; Sylva from Austria; Sector from Sweden; and PEP I and II from Russia. Several of these European saws were obvious copies of the Wolf saw, even down to such items as chain rivet dimensions. However, none of the sawing mechanisms proved as satisfactory as that of the Wolf saw."

—Indiana forester Charles Miller

Charles Miller goes on to write that there was one aspect in which the Europeans came out ahead: "The saws were all portable, and ranged in weight from 60 to 120 pounds [27–54 kg]. At that time the Europeans excelled in production of small gas engines, and as one might expect, these foreign saws were all powered by that type of motor."

Das neue Modell der
„Rinco"-
Baumfäll-
und Ablängmaschine
in höchster Vollkommenheit und Zuverlässigkeit ist für den Wintereinschlag
für Sie unentbehrlich.
Nachweislich 10 facher Zeitgewinn — 80 proz. Lohnersparnis. Glänzende Gutachten
staatl. und städt. Forstbetriebe.
A. Stihl, Stuttgart, Rotebühlstr. 43.

E. Ring Co. of Berlin produced the first fully portable gas chainsaw by 1925.

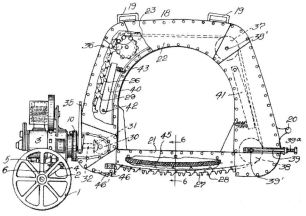

Arsneau of Minneapolis patented the Bow-Frame Chain Saw in 1925, but its wheeled frame limited portability.

The Arsneau didn't impress BC loggers. Model 2 seen here.

For decades, the German manufacturers Stihl and Dolmar have run neck and neck in claiming to be the first inventors of the gasoline-powered chainsaw. They both came up with their first models in 1927. However, the German company Gerber produced a portable two-cycle-engine-powered bandsaw in 1919. Although little information remains, the Gerber must have been a hard saw to handle, with its heavy engine mounted at the top of a triangular frame, which doubled as a bow guide.

By 1925, the E. Ring Company of Berlin had produced both the Erco and the Rinco chainsaws, and collectors have found separate Rinco advertisements that respectively name Andreas Stihl and Dolmar founder Emil Lerp as sales agents, Stihl in Stuttgart and Lerp in Hamburg. Therefore it seems that credit for the first gas chainsaw should go to the short-lived Sector saw of Sweden and for the first fully portable gasoline chainsaw to E. Ring Co. of Germany. The Rinco had all the features necessary to allow it to be used in forestry work but the cost was prohibitive for the ordinary forest worker of the day.

In 1925 the Arsneau & Sons Saw Company of Minneapolis patented the Bow Frame Chain Saw. Later bow saws were designed with a thinner bar intended to guide the chain through a cut without getting pinched, but this doesn't seem to have been the idea with the Arsneau, since pictures show its bar to have been a good six inches wide. The whole contraption was designed to be mounted on a wheeled frame anyway, reducing any finer concerns about minimizing weight, so the bow would certainly have added stability to the cut—and safety as well, since the chain was covered by a stout metal housing ringed with handles that made it possible for more than one person at once to get a good grip on it. This was necessary because the bow frame and cutting assembly were awkward to handle, especially when turned sideways in "felling position," where three men were required to guide the cutting chain through the tree.

The Arsneau also featured a "hand-cranked lift or elevator to raise and lower the machine vertically." This could raise the bow frame when, for example, it was used to buck big logs, or lower it in felling position to make a cut nearly flush with the ground. Mounted on a pair of wheels and powered by a 2.5-hp motorcycle engine, at 150 pounds (68 kg) the Arsneau was lightweight compared to some. Although the Arsneau caused quite a stir when it was demonstrated at the McGoldrick Lumber Company's sawmill in Spokane, Washington and on the Vancouver Court House lawn for the BC Loggers Association, there was considerable scepticism about its practicality in the woods.

I find as early as April 1927 our Association recorded in its minutes that it had a proposal from a man named Arsneau to demonstrate a power saw. Arsneau brought up a power saw mounted on two wheels and we placed a 30-inch log on the Court House lawn where a demonstration was held which showed that the saw would cut if it could only be gotten to the timber. No one could figure out how we could drag a saw on wheels over rough ground and through underbrush, so that particular saw was forgotten for the time being.

—John N. Burke, former secretary of the BC Loggers Association

The story of this Arsneau saw was a common one in this age of prototypes: once again a working saw had been designed, but was kept from general application by its sheer size. Weight was the critical factor, and new developments in welding and casting lightweight metals such as aluminum, and the even lighter magnesium, put Europeans on the cutting edge of innovation during the 1930s. Stihl and Dolmar were putting out saws of comparable weight and quality and promoting them as best they could to the worldwide market.

By 1929 Stihl was selling 100 saws a month. *BC Archives, NA 07137*

Andreas Stihl and the German Invasion

Andreas Stihl, born in 1896 in Zurich, Switzerland, studied mechanical engineering and by his late twenties was a busy mechanic, repairman and inventor in Stuttgart. The

young Andreas had a fertile and inventive mind—in 1923 he built a gasoline-powered washing machine—but he made his living in the lumber industry. Stuttgart is in the southwest corner of Germany, close to the French and Swiss borders and on the edge of the Black Forest, and Stihl did sales and service for a local mill and industrial supply house.

Among his clientele were Black Forest loggers, who were seeing every facet of their business mechanized except the backbreaking process of felling and bucking. Hired to service one of the von Westfelt Sector saws, Andreas managed to get it running, but was dismayed that his customers had to depend on what he felt was a heavy, inefficient and unreliable machine.

For a time, while he acted as a sales agent for Rinco saws, Stihl took a long hard look at the new machines coming out of Sweden, the USA and his own country. Power saws, Stihl observed, were getting bigger and heavier when they should have been getting smaller and lighter. In his home workshop, Stihl developed an electric model to be operated by two men. It weighed 140 pounds (64 kg), but still it was a portable saw, even though its 3-kilowatt motor needed an external power source—either the nearest land line or a portable generator. Stihl tested his prototype in the Black Forest and sold it immediately.

Far left: Andreas Stihl.

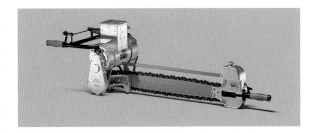

Stihl electric Model TFM , 1926. *Courtesy Andreas Stihl-Waiblingen, Germany*

35

By 1926, no longer a Rinco sales rep, he founded "Maschinenfabrik A. Stihl." The new company's principal business was the manufacture of accessories for steam boilers, but that was soon outpaced by the demand for Stihl's saws. Within a year Stihl had hired Karl Gutjahr to take over the hands-on work of manufacturing the saws, and had moved Maschinenfabrik A. Stihl into new quarters in Bad Constatt, a Stuttgart suburb.

Soon the company was enjoying brisk sales of its electric 3JW throughout Germany as well as to lumberyards and builders in Holland, Belgium, France and Switzerland. Stihl also sold some saws overseas—in the US the cost was $485 for a unit with a 5-foot (1.5-m) bar. But logging companies were slow to adopt the new saws—inhibited by their high price, by the difficulties of providing an electrical power source in the woods, by the fact the voltages and hertz of electricity in Europe were different than those used in North America, and by the sales resistance that the first chainsaws encountered all over the world. Electric motors and internal combustion engines were still very new technologies, and operators were intimidated firstly by the learning curve they would need to undertake to operate the machines, and secondly by the much more formidable task of learning to maintain and repair them.

Last but not least, forest workers felt that powered saws would put them out of work. The Depression had seized Germany as it had the rest of the western world, and the last thing the workers needed was a further threat to their employment. It was not uncommon for Stihl's salesmen to encounter surly sales resistance and outright belligerence in the course of their travels.

Stihl's sales continued in the face of all setbacks, but the company also continuously sought methods of making a better and more portable chainsaw. The passion of the day

Top and above: Stihl quality control. *Courtesy Andreas Stihl-Waiblingen, Germany*

Below: The Stihl Model B. *Andreas Stihl-Waiblingen, Germany*

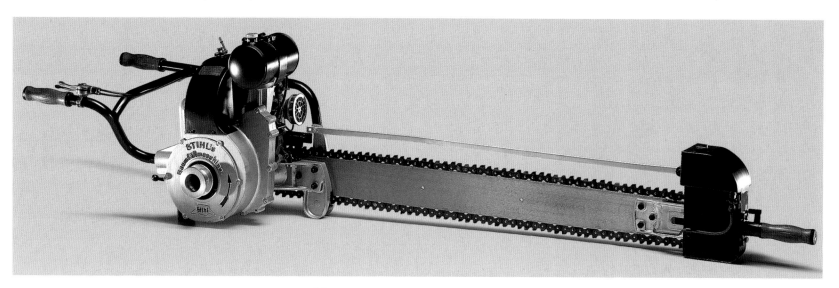

was racing cars, and new alloys and other technical developments were making their engines not only faster and more powerful, but lighter and more durable. Stihl saw no reason that these advances couldn't be put to work in the woods. In 1927, his first gasoline-powered saw was not quite a race car. It was so large that the bar and chain assembly had to be detached from the engine in order for it to be moved, but it was a beginning. In 1929, Stihl incorporated a two-stroke DKW motorcycle engine into the 6-hp, 101-pound (46-kg) two-man Model A, with a newly designed friction-reducing guide bar.

Stihl experienced great expansion through the late 1920s and 1930s. By 1929, the company was selling one hundred saws a month, and had set up a separate sales office in northern Germany. In 1930, Andreas Stihl attracted crowds at the Leipzig European Trade Fair with a saw that, at 7.5 hp and a mere 127 pounds (58 kg), he called "The Midget." The excitement generated by this machine brought in more orders from North America and the USSR.

Andreas Stihl was a charming, persuasive man with an infectious belief in his new product. In 1931 he made his first sales trip outside of Germany. Maschinenfabrik A. Stihl got a major boost when he returned from the Soviet Union with a single order for several hundred saws. Over the years he managed to work out solutions to the problems not only of manufacturing and exporting saws, but of training technicians from his client countries in their use, maintenance and repair.

In 1935 Stihl introduced an automatic chain oiler to relieve busy saw operators of the need to keep squirting lubricant on the chain. His new saws also featured the centrifugal clutch, which engaged the chain automatically when the engine was revved up and disengaged when it slowed down, a great improvement over the manual clutch (a similar design was patented in the USA in the same period, but Stihl was the first to use it on their saws). A year later Stihl modified the existing two-strap chain into the three-strap design that is the industry standard to this day. Also in 1936 Stihl issued its first bow saw, the BBU, designed for mill work rather than for use in the bush.

As falling and bucking steadily became mechanized, Andreas Stihl feared that his German competitor, Emil Lerp's Dolmar, would become the brand of choice in the booming Canadian forest industry. And when he heard that a Dolmar saw had been tested in British Columbia, Stihl realized it was time for drastic action. In autumn 1937 he boarded an ocean liner for Montreal, where he set out across the continent with the five sample saws on a mission to establish Stihl as the chainsaw of choice in North America. Working tirelessly and taking full advantage of his fabled European charm, he was able to woo major clients like the BC timber giant Bloedel, Stewart and Welch and set up distributorships to service them. It was a great start that seemed to open the door for a dominant role in global markets, but there were to be a few bumps along the way.

The 1926 Stihl electric sold for $485 in the US. *Courtesy Andreas Stihl-Waiblingen, Germany*

Difficulties of providing power kept the early electrics out of the woods. *Courtesy Andreas Stihl-Waiblingen, Germany*

Model BD was the first Stihl with enough power for use in big
timber. Tested and improved, it evolved into the model BDK
(above). The Model BK (below) was used for bucking.

Collector Marshall Trover, photo Brian Morris

This Stihl BK prototype is beautifully restored.
Collector Marshall Trover, photo Brian Morris

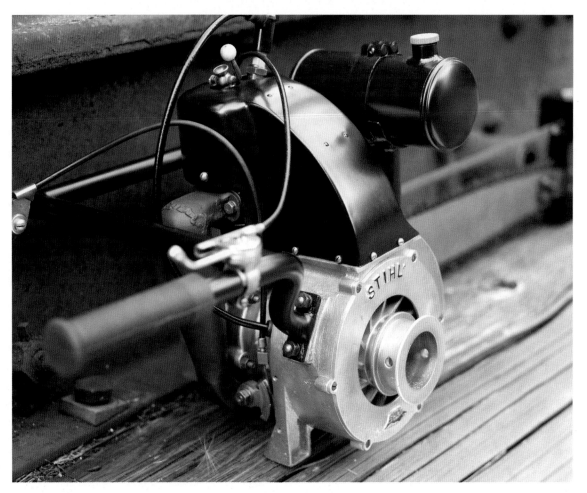

Top left: The Stihl BK used aluminum castings, reflecting Andeas Stihl's belief chainsaws should be getting smaller and lighter. *Collector Marshall Trover, photo Brian Morris*

Left: Strap starters were the order of the day. *Collector Marshall Trover, photo Brian Morris*

Top right: At 22 kg (48.5 lb), the BK was a featherweight in its day. *Collector Marshall Trover, photo Brian Morris*

Above: The head ends or "helper handles" of the early two-man saws were cumbersome affairs with their own oil reservoirs. *Collector Marshall Trover, photo Brian Morris*

41

This Stihl BDK was produced in 1937 and sold by Mill and Mine Supply to Polson logging in Hoquiam, WA.

Collector Marshall Trover, photos Brian Morris

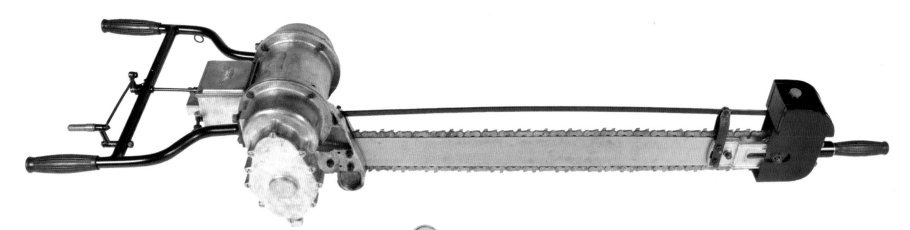

Above: 1938 Stihl STD. Stihl electric models were produced for decades in various versions. *Collector Marshall Trover, photo Brian Morris*

Right: This awkward-looking 1941 Stihl Type 41 electric bow saw was used in construction and lumber mills in Europe. *Collector Marshall Trover, photo Brian Morris*

Below: Stihl BDKH. *Collector Marshall Trover, photo Brian Morris*

This 1938 Stihl BDKH never made it to North America but evolved into the KS43, which became the primary German saw of WWII and was produced in such numbers that Stihl enlisted, among others, Dolmar in Germany and PPK in France to manufacture clones. It was an advanced design for the time with many new features such as an automatic rewind starter, gas and oil tanks built into a unified magnesium cast housing, and was lighter than wartime models built in North America.

Collector Marshall Trover, photos Brian Morris

DOLMAR
MOTOR-
SÄGEN

REIN DEUTSCHES
ERZEUGNIS

Far right, top: German Inventor Emil Lerp, founder of Dolmar.

Far right, bottom: Alfonso Lange, Lerp's business partner after 1929.

The first Dolmar Model A was a hulking monster with a 246-cc engine.

DOLMAR

Dolmar: The Other German Giant

When Stihl had been planning his first chainsaw with a gas engine, another German was already testing just such a machine. By 1927 Emil Lerp had, like Andreas Stihl, acted as a sales agent for Rinco chainsaws then gone ahead and built his own. Although Stihl's first saw was an electric model, the first of Lerp's chainsaws was powered by a huge 15-cubic-inch (246-cc) two-cycle, air-cooled gasoline engine.

The saw met with some resistance in Lerp's home province of Thuringia. For one thing, if we think a large modern chainsaw, with its 90-cc motor, die-cast lightweight parts and muffler is noisy, think what Lerp's saw must have been like, with an engine almost three times the size. In most parts of the world, even as late as the 1920s, the internal combustion engine was in itself an upsetting new sound. No one was prepared for the sheer awful amount of noise these things made. Loggers made the point that the noise of the saw alienated woodworkers from each other and from the sounds of the forest around and above them. This was only one among a host of objections to the new machine.

Lerp, however, finally managed to have his machine tested on the heavily wooded slopes of Thuringia's Mount Dolmar. The tests were so successful that he bestowed the mountain's name on his new saw and the company that was to produce it. In 1928, the German forestry magazine *Forstarchiv* reported on tests of a new Dolmar: "There were no major interruptions. The engine worked faultlessly." However the magazine's critics expressed reservations. "The economic use of the Dolmar will have to be proved by more tests."

In 1929, Lerp entered into partnership with Alfonso Lange. By 1930, their Dolmar Maschinenfabrik in Hamburg had two hundred employees. Throughout the 1930s, Dolmar concentrated on making lighter and more powerful saws, and on marketing them inside and outside Germany. The model C, which the company produced from 1930 to 1937, offered sophisticated features such as a swivelling bar so the saw could cut vertically or horizontally with the engine sitting upright, a laminated guide bar featuring replaceable steel "wearing strips" and a gravity-feed chain oiler.

By 1937 Maschinenfabrik A. Stihl was running neck and neck with Dolmar for the position of Europe's leading chainsaw maker, but the companies had still only scratched the surface of the potentially huge North American market. Dolmar had set up a distributor in New York in 1934, and in 1936 CEO Emil Lerp had spent a summer demonstrating his C1 in the pulp woods of northern Quebec.

At least two Canadian Pulp and Paper Association members, Price Brothers & Company in Chicoutimi and the Canada Paper Company in Windsor Mills, purchased Dolmars and performed serious test runs on them early in the fall of 1937. They were impressed with the C1's usefulness in the mill yard, but didn't find it suitable for the more demanding work in the bush. The saw was tested on pulpwood balsam and spruce trees no more than nine inches thick, felling them and bucking them into four-foot lengths.

The work demanded a lot of saw-handling, and at 96 pounds (44 kg), the C1 was just too heavy. "The weight of the saw, and the difficulty with which it is handled in slash and underbrush, prevents its effective use at the stump," reported one observer. There were also starting problems: although they found the Dolmar to be rugged and "mechanically perfect... some difficulty was experienced in starting the motor at certain times (on one occasion it required twenty-two minutes), but this was attributed to the continuous cold and wet weather." Nevertheless, Dolmar's increasing sales through the 1930s gave the company the capital it needed to keep researching and developing better chainsaws.

This kind of persistence soon gained both Dolmar and Stihl a solid foothold in the North American market. For a few years the company enjoyed a good relationship with its North American distributors and customers. During those years, however, political forces were undermining German trade with the rest of the world.

Above: Sharpening chain in the field was a two-man affair.

Left: Falling low on the stump with a Dolmar Model CK.

47

Above: With the Dolmar CL's swivelling bar it was easy to make diagonal undercuts.

Above right: Bucking a hardwood butt with an electric Dolmar.

Right: Dolmar actively pursued overseas markets.

Opposite page: Quality-control inspection of a shipment of Dolmar CLs, Hamburg c. 1949.

The 1933 Dolmar C influenced the designs of many new manufacturers outside Germany as the world went to war in 1939. *Collector Marshall Trover, photo Brian Morris*

With its big 200-cc JLO motorcycle engine, the 1930-37 Dolmar C weighed in at 45 kg (99 lb). Like most saws of its day, it was configured for two-man use only. The operator had to keep both hands on the handlebars and there was no front handgrip at all. *Collector Marshall Trover, photos Brian Morris*

Logging in the Pacific Northwest

Despite the fact that the business of logging has gone mechanical with a subsequent increase of efficiency and a decrease of man power, the cross cut [saw] has been able to hold its own with undeniable superiority. It may be said that the hand cutters appeared to be getting nowhere as they whipped their blade, the kerf moving forward at a snail's pace. Yet their transit from tree to tree with only the saw, axe and wedges to encumber them recaptured many of the lost moments. The power saws found their greatest difficulty in transportation, thereby establishing the two major marks for the manufacturers to shoot at—less weight, greater mobility.

—Buck Weaver, *Timberman Magazine*, 1935

The need for a practical, truly portable chainsaw was nowhere more acute than in the Pacific Northwest. The temperate rainforest of the British Columbia and Washington coasts boasted not only some of the world's biggest trees, but some of its most rugged terrain. By the 1920s, much of the level ground—what little of it there was to begin with—had been logged off. Logging operations were moving farther back into the mountains to cut on more difficult terrain.

At the same time, forest companies were pressing for ever-increasing production. Higher production, lower cost was the rationale behind mechanization, and other facets of the industry were being rapidly mechanized. The bull team had given way to the railroad, which was fed by high-lead yarders, all powered by steam. It was the age of steam. Railroad logging camps grew to the size of small towns. Coastal valleys could be swept clear of fir, cedar and hemlock on a scale undreamed of by previous generations. Yet the actual felling of trees and bucking them into manageable lengths was still done with crosscut saws and axes.

Aided by steam-powered equipment like this yarder, timber production doubled in the Pacific Northwest between 1912 and 1930, making it one of the world's leading timber-cutting areas. *University of Washington Special Libraries, Special Collections CKK01624, photo Clark Kinsey*

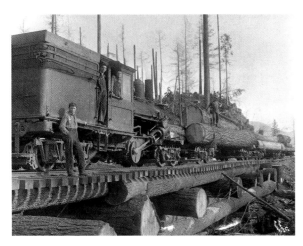

Steam train loaded with logs, Snohomish, WA, 1913. Every facet of logging was being mechanized except falling. *University of Washington Special Libraries, Special Collections PIC0115, photo Lee Picket*

Martin Fossum

During the handfalling days mostly Scandinavian and central Europeans were involved in the falling—very few Canadians. They were used to hard work back there. I hate to say it but I know when I was in charge we tried out Canadians. We had a class up in camp and I had an instructor for each group. In spite of the fact those boys were young fellows and very husky, only a very small percentage turned out. I'd say only about five per cent carried on after graduation. They just gave it up. Too much work, too cumbersome. But the Scandinavians, they're used to pulling the Swede fiddles (crosscut saws). They didn't like to run a power saw.

—Martin Fossum, ex-faller

CHAINSAWS

A set of Pacific coast handfallers pose beside a first-growth Douglas fir with tools of their trade. They first inserted metal-shoed planks called springboards into the trunk of the tree to get up into the clear wood above the butt swell, then jumped up and chopped the undercut with special narrow-headed falling axes before completing the job with their crosscut saw. It was hard work but the men refined it to an art and held off mechanization longer than in any other facet of logging. *University of Washington Special Libraries, Special Collections PIC0115, photo Lee Picket*

Above: Brute strength wasn't everything in a handfaller; some of the best were light but skilful Japanese. *University of British Columbia, Rare Books and Special Collections, Japanese Canadian Photo Collection, image # XXVII-21*

Right: Less glamorous and even more laborious than falling was the job of bucking the big timber into moveable lengths, usually done by a single worker. *University of British Columbia, Rare Books and Special Collections, image # BC 1456/62*

Mind you, the handfallers did a good job. As Olaf Fedje noted, when the first Stihls came into the woods, they only cut 30 percent more than the handfallers, while the handfallers' quickness moving from tree to tree tended to equalize production. Given the added cost of buying and maintaining the power saws, it is easy to see why handfallers held off mechanization long past other parts of the logging process. Still, between 1912 and 1930, timber production in British Columbia had doubled, from 6.7 million to 12.7 million cubic metres. During the 1940s it would double again. To men who could see where the forest industry was going, it was obvious that falling and bucking would have to be mechanized eventually.

In the mid-1930s Sidney Smith, general manager of BC's largest lumber outfit, Bloedel, Stewart and Welch, got together with the company's supervisor of falling and bucking, Jack Challenger, to find solutions to the problem. At first, they looked at the Wolf saw from Oregon, which by that time was available in electric, pneumatic, or experimental gas-driven models, and the gas-powered saw from Dow Pump & Diesel Engine Company of Alameda, California.

The Wolf saw didn't seem rugged enough for the big timber they were tackling on the coast. In 1936 Challenger imported a Dow saw to use at BS&W's logging operation at Great Central Lake on Vancouver Island. At a glance, the Dow looks less like a chainsaw than a piece of antique field artillery. It was powered by a four-stroke 18-hp V-2 Indian motorcycle engine, weighed 460 pounds, and was mounted on wheels adapted from a World War I fighter plane. It sold for $990, a huge amount of money in those Depression years. Although a powerful wood-cutting machine, aggressively promoted by its Alameda manufacturer and widely tested, the Dow was not used for very long.

BC Loggers felt the same way about the Dow that they did about the Arsneau. The problem was moving it through the woods. The Brown brothers, Alvin and Gunny, ran the Dow saw for BS&W at Great Central Lake in 1936. As the story goes, one day they simply left the Dow in the woods, returned to camp and went back to hand-sawing. Over fifty years later, Gunny told Jack's nephew, the chainsaw buff Dave Challenger, his still-vivid memories of the Dow saw: "If you could get that outfit to a tree, it would cut to beat hell."

The camp salvaged the Browns' abandoned Dow and converted it into a fire pump. Meanwhile, Smith and Challenger examined other options. What was needed was a gas-powered saw weighing no more than 125 pounds so that two strong men could get it through the bush. It had to be dependable and it had to be simple enough that it could be fixed in the field by mechanics with a moderate amount of special training. Above all, of course, it had to beat the competition: "The saw must be of such an order that it could maintain a consistently higher output per man than obtainable by the highest skilled hand faller."

Throughout 1935 and 1936 Jack Challenger travelled throughout the Pacific Northwest, corresponding with and seeking out colleagues with similar interests. He met with Mark Forrest of the US Forest Service in Montana, and traded notes on Dow saws, which had disappointed BS&W on Vancouver Island, and on Wolf and Reed-Prentice saws, which the Montana USFS had tried and found wanting.

At the same time, the BC Loggers' Association had heard about a German outfit who claimed to make a lightweight, powerful two-man chainsaw. The German outfit was of course Maschinenfabrik A. Stihl. When Andreas Stihl shipped his first Canadian chainsaw overseas in December 1936, the BCLA sent the saw on to the BS&W camp at Franklin River on Vancouver Island.

The Stihl BD was simpler than the other chainsaws on the market, driven by a single-cylinder, two-cycle, high speed engine, easier to handle, and lighter—with a 60-inch (152-cm) bar, it weighed a mere 120 pounds (54.5 kg). Manufactured from 1934 to 1939, the BD's weight of course varied according to the length of the bar.

In the hunt for a practical falling machine, west coast loggers tried the Atkins electric chainsaw, but the challenge of slinging heavy electric cables through the bush proved prohibitive.
British Columbia Forest Service, photo G.M. Abernathy

Because of the helper handle on the end of the bar, two-man saws had to be longer than the tree was broad. When they weren't, fallers like these had to carve sections out of the tree.
University of British Columbia Library, Rare Books and Special Collections, image # BC 1930/228/2

Top and above: The 1933 wheel-mounted Dow Low Stump Power Saw worked fine as long as the ground was level and open.

The Dow used an 18-hp Indian motorcycle engine with a model A Ford rear end for a transmission and a hand-cranked hoist for raising and lowering the bar. It weighed 460 lbs (209 kg) and cost $990 in 1933.

Collector Marshall Trover, photos Brian Morris

The Dow "cut like hell"—if you could get it to a tree. *Collector Marshall Trover, photo Brian Morris*

Chainsaws were great for bucking, but the early ones all required two men and bucking was traditionally a one-man job. Here Olaf Fedje tests a two-man Shade Model C that has been modified for one-man operation.

In Europe, the BD was a success in meeting local demands. But those demands were for cutting smaller trees on a smaller scale than in BC coastal logging operations, where the BD was not quite up to the task. Jack Challenger wrote, "The machine operated for a while and broke down for a while as one weakness after another developed."

However, Andreas Stihl could not have found a better testing ground for his fledgling saws. Communities such as Franklin River were isolated but they were by no means resourceless. To service the huge amount of machinery that was being devoted to logging, they had machine shops and machinists who were experienced at tackling new problems and being forced to handle them on their own. As Dave Challenger said almost seventy years later, "They made everything right there. They had big railroad shops; they had lathes that were twenty feet long; they could make anything at all." As much as they could, BS&W's machinists redesigned, repaired and rebuilt the Stihl, and Challenger kept the German manufacturer posted as to what was going wrong and how they were trying to remedy it. Stihl did his best to solve the problems posed by Franklin River, and whenever he had found an effective solution, he had patented it. By the late 1930s, he had copyrights on most of the essential features of the modern chainsaw—the guide bar, the automatic oiler, the centrifugal clutch and the compression release, to name a few.

To follow up on these early contacts, Andreas Stihl journeyed to the New World and crossed the continent in July 1937, bringing with him samples of his latest, improved design. As he headed west, Stihl looked up logging outfits, stopped in and demonstrated his saws. The logging community was excited—so excited that by the time Andreas arrived in Vancouver he'd sold every one of his sample saws. Andreas proudly showed them his brochures and boasted about the wonderful reception his saws had received as he crossed the continent, but the BC lumbermen brushed aside the brochures and demanded to see the real deal.

Andreas Stihl knew he wouldn't be passing this way again soon. If he couldn't sell saws while he was meeting the west coast lumbermen face to face in downtown Vancouver, bringing to bear the full force of the endorsements of experts, his carefully composed brochures and his European charm, he sure wouldn't be able to do it from Bad Canstatt. Several more saws were on their way to him from the factory in Germany, but there was no telling when they would arrive.

Stihl spent the best part of the next week retracing the 2,200 kilometres (1,400 miles) back to Winnipeg. There, he looked up his latest customer and managed to borrow back the chainsaw he had just sold him. Then he turned around and once again headed west to Vancouver. With a working model to show, he took a wad of advance orders, visited BS&W's Franklin River camp on Vancouver Island, and signed up the fledgling D.J. Smith Company of Vancouver as his North American distributor and sales agent.

The German manufacturer's hands-on marketing methods paid dividends. By 1939, Bloedel, Stewart and Welch were using thirty Stihl machines in their logging operations, and had established themselves as pioneers in mechanized falling and bucking. Other companies, large and small, were staying with handfalling—but they were also staying with the problems of handfalling, and with the problem of finding fallers.

At BS&W the saws eased the problem. Unpredictable and dangerous, they spewed smoke, roared like machine guns and were hard to handle—but for many younger loggers it was love at first sight. The trouble was, they were being called off to war.

It Was a New Life

The thing people said about the power saws at first was that they couldn't stand the noise, the noise was so terrible. I believe there was some truth to that. Hand falling was so quiet. All of a sudden this terrible monster came along and the noise bothered them to start with. I guess something we should have learned earlier was to get ear muffs. A lot of people lost their hearing because of that noise.

The danger part of it? They were afraid of them because of the noise, I think primarily. At that time the mufflers didn't cut the noise down, they were very loud. It took some time to get the new people used to it, to that terrible noise. But it was lighter work. It wasn't the heavy slogging. The younger people took to it quite readily. But the old timers fought it tooth and nail. They didn't think it would ever take the place of hand saws. But productivity in those days, the late thirties, was important. Hand fallers would fall about 7,000 board feet per day. With the advent of the first power saws, the Stihls from Germany, production was about 10,000 board feet per day, which is quite an improvement. Considering the saws at that time were 130 or 140 pounds.

There were people who quit the woods because they wouldn't pack that monster around. It was a new life. If we wouldn't have trained a lot of new fellows it would have taken a lot longer to get it going.

—Olaf Fedje, a founder of Fedje & Gunderson, BC falling contractors, interviewed by Ken Drushka, *Working in the Woods*.

Top: The old scratcher chains took a lot of filing, making the saw filer an indispensable man in early camps. *University of British Columbia Library, Rare Books and Special Collections, image # BC 1930/550*

Above: Power-filing scratch chain took a light touch on the emery wheel. *University of British Columbia Library, Rare Books and Special Collections, image # BC 1930/188/13*

World War II and Its Effects

World War II and the World Chainsaw Industry

T he war had a huge effect on the chainsaw industry, bolstering some manufacturers with military contracts, and hindering others who found that their parts sources had been co-opted for wartime needs. In Allied countries the most important effect was in removing German competition, while at the same time nullifying Germany's international copyrights, opening the way for anyone who wanted to make use of the innovations patented by Stihl and Dolmar.

This beautifully restored 1945 Teles was produced for the British Army in WWII. It was similar to the Danarm of the same period.
Collector Marshall Trover, photo Brian Morris

The Stihl KS43 was the official saw of the German army during WW II. *Collector Marshall Trover, photo Brian Morris*

Wartime was good and bad for A. Stihl Maschinenfabrik. First it brought thousands of orders for the KS43, then it caused the Stihl factory to be designated a supplier of strategic materials and bombed to oblivion by the Allies. *Collector Marshall Trover, photo Brian Morris*

Stihl During World War II

The subsequent loss of its overseas contacts due to the war did not actually threaten the Stihl company's existence. On the contrary, armies all over the world had immediately recognized the value of chainsaws for building roads, or obstructing roads when strategy called for it. The German forces needed chainsaws and lots of them. Stihl innovations in this time began with magnesium castings and chrome-plated cylinders in the 1930s, and continued through the war years, introducing die-cast housings in 1943. Hitler's forces enjoyed state-of-the-art chainsaw technology with the two-man Stihl KS43—they demanded so many of them that a factory in occupied France was turned over to producing clones of these saws.

Stihl's contribution to the Nazi war effort did not go unnoticed by the other side. In 1944, the factory and head office at Bad Canstatt were bombed and totally destroyed by Allied bombing raids. As Germany began to take on most of the western world, first Stihl had lost the patent protection afforded by international copyright laws. Since 1938, copies of Stihl saws became the basis for new models of chainsaws that had appeared all over the planet. With no more international trade, copyrights, machine works or factories, by the time the war ended in 1945 Stihl appeared to have lost everything.

With die-cast magnesium housings and recoil starter, the KS43 had very modern features for the time. *Collector Marshall Trover, photo Brian Morris*

Above: Shade Forest King. *Collector Marshall Trover, photo Brian Morris*

Below: Burnett B29. *Collector Marshall Trover, photo Brian Morris*

Burnett

Shade Engineering/Burnett

In the midst of the World War II chainsaw boom, Bloedel, Stewart and Welch of British Columbia were one of the few companies big enough to go their own way in finding a new source for saws.

In the late 1930s BS&W had sent one of their Franklin River machinists, R.W. (Bob) Shade, to the Stihl plant in Stuttgart to see how Stihl saws were made, and to discuss their performance in the west coast woods. Upon his return Shade put to good use what he had learned. In 1940 BS&W put up part of the money to establish Shade Engineering on Granville Island on Vancouver's False Creek. The business built parts for Stihl saws, but by November 1942 BS&W had decided that parts were not enough. They bought out Shade's share and made him manager of Shade Engineering Works Ltd., a new BS&W subsidiary established solely to provide them with saws. For the next year and a half, working under directors Jack Challenger and Prentice Bloedel, Bob Shade imported one-cylinder Villiers engines from England and built the Shade Model B, a relatively lightweight, 250-cc two-man falling saw and the Model C, a 125-cc bucking saw. The saws were branded Forest King, and proved to be more than satisfactory for BS&W's requirements.

Above: A BC faller using a Burnett Forest King, one of several copies of the Stihl BDK that appeared as war halted exports from Germany.

Below: Bucking with a modified two-man Shade Model C.

NEWS about the NEW B-30

Burnett POWER CHAIN SAW

The success to this point with the Forest King brought BS&W to a crisis in their career as chainsaw manufacturers. The demand for the Forest King was becoming widespread, but the sixty units they had produced were enough for their present needs. For the time being, then, Shade Engineering Works was only good to them as a source of parts—unless they wanted to diversify their operations and enter the chainsaw manufacturer's domain. Reluctant to divert their attention from the booming forest industry, the company made it known in Vancouver manufacturing circles that they were getting out of the chainsaw business.

At this point a new player entered the game. A well-known car and motorcycle dealer, Fred Deeley, was also the Vancouver sales agent for Villiers, and he didn't want to see the demise of the Forest King. He bought out BS&W's share in the machine works, retaining R.W. Shade as manager. They then began to manufacture and market the Forest King for general sale. In November 1944, Deeley turned around and sold the company to R.M. (Dick) Burnett.

Since the success of the Forest King, Shade Engineering had not been generating the new initiatives that a growing business needs to stay solvent. Dick Burnett consulted with Jack Challenger to design a new saw, the B-6, built around the single-cylinder Villiers. The new two-man B-6 (125 pounds/57 kg with a 48-inch/122-cm bar) debuted in February 1945. In the same month, with the war still raging, R.W. Shade, who was still on the payroll, was called up for service. In March the company became Burnett Power Saws & Engineering Co. Ltd.

By the end of the summer, five B-6 models were being field-tested, and in October they began manufacturing six per month. Early in 1946, new shipments of Villiers engines went into the Burnett B-29, which really established the former Shade works as a force in the chainsaw business. They also introduced the 125-cc Burnett Model C, and by late in the year Dick Burnett was able to boast:

We supply machines and do all the repair work for the Bloedel Stewart & Welch camps situated at Franklin River, Great Central Lake, Sproat Lake and Bloedel, BC. This company now has approximately 60 of our smaller model and 47 Model B-29s. We will, we believe, install an additional 25 machines in the Bloedel Co. as soon as machines are available. We have 13 machines at Victoria Lumber Co., we have 13 machines at Northwest Bay, both of these companies being owned or controlled by H.R. MacMillan Co. We have 13 machines at the Englewood Division of the

Above: Power falling crews often consisted of a young fellow with strong arms at the controls and an old-timer with know-how at the tail end. *British Columbia Archives, E-03397*

Below: New equipment called for new falling methods.

Above: Shade Forest King. *Collector Marshall Trover, photo Brian Morris*

Below: Burnett B29. *Collector Marshall Trover, photo Brian Morris*

Canadian Forest Products Ltd., and 7 machines at their Harrison Lake Spring Creek operation. In addition individual contractors have 5 of our machines at Elk River Timber Co. The remainder of our machines are spread over Vancouver Island and the Lower Mainland.

BS&W continued to be Burnett's best customer. Under Jack Challenger's guidance, they hired young men and introduced them to the new saws as part of an experienced crew. Then they apprenticed them with expert head fallers. During the war years, as BS&W added between 250 and 300 power-saw fallers to their work force, the rest of the forest industry followed the trend.

Mel Parker

"In those days when they started with the power saws they had a school down here at Nanaimo. A guy by the name of Ole Buck started it. And they'd train these guys on the saws. He'd run the saw and they'd put a faller with him, probably a handfaller. He's on the light end, the faller. He took care of the falling of the tree and the young guy just ran the machine. He was the second faller, or the machine man. That's the way they started out. Then they got down to one-man saws. Now it's every man for himself."

—Mel Parker

During WW II Shade Engineering began producing the Forest King, a copy of the Stihl BDK. *Collector Marshall Trover, photo Brian Morris*

Above and left: In 1944 Shade Engineering was taken over by R.M. (Dick) Burnett, who began manufacturing the Burnett B-29 at the rate of one or two per month. *Collector Marshall Trover, photos Brian Morris*

D.J. Smith

Old versus new in 1940: power fallers at Wellburn Timber tackle cedar with a D.J. Smith Timberhog while handfallers look on, their crosscuts idle. *Courtesy Steelworkers local 1-80*

D.J. Smith

In 1936 Don J. Smith, Bill Vaughn and others assembled in the basement of a Vancouver hotel for a meeting that would change the course of chainsaw history. Their immediate purpose was to start a distribution and service business aimed at the forest industry, but the meeting sowed seeds for no less than the North American expansion of Stihl, the Canadian construction of Reed-Prentice's Timberhogs, and the creation of Hornet, IEL Beaver and the Pioneer saw. During Andreas Stihl's visit in 1937, D.J. Smith signed a contract with Stihl appointing him as the sole sales agent for Stihl products for all of North America. With special orders sometimes taking months to arrive in the following years, Smith and his partners began to convert existing parts and then to make their own.

In the process, they became expert at every aspect of saw production. Incorporating as D.J. Smith Equipment Ltd., they moved to larger quarters, and when war broke out in 1939, they had everything they needed to start producing their own chainsaw. When Smith was informed that the freighter carrying their latest shipment of Stihls had been impounded by the US government at the New York waterfront, he immediately rushed into production the D.J. Smith Model A, as close to an exact copy of the Stihl Model BDK as their machine shop could manage.

But in the early years of the war, Reed-Prentice, the big US tool manufacturer who had produced chainsaws for Wolf, made D.J. Smith an offer the partners couldn't refuse, proposing to finance the construction of its chainsaws in Vancouver. From 1940 to 1942, Don

Above and right: When WWII cut off supply of saws from Germany in 1939, Stihl's Vancouver distributor, D.J. Smith, started making his own saw. The D.J. Smith Model A was as close to an exact copy of the latest Stihl as D.J. Smith could make it. *Collector Marshall Trover, photos Brian Morris*

J. Smith himself continued business as usual, managing a staff of about sixty people at the business he had founded, but doing it under the Reed-Prentice name. Whatever its name, the business produced some very successful saws, including the D.J. Smith Models A and D. Their most famous saw, Model K, was better known as the Timberhog, whether it was produced under the D.J. Smith name or by Reed-Prentice. With a 60-inch (1.5 metre) bar, the Timberhog/K weighed 150 pounds (68 kg).

Although the technology was accelerating steadily, enabling saws to be made smaller and lighter, the west coast still had lots of big trees and the Timberhogs were big saws. Before spring-loaded starter recoils, the starters were simple pulleys mounted on the flywheel that had to be rewound by hand. Years later, Roy Byers recalled, "All were wicked to start and seemed to possess minds of their own. If the operator was not careful in the positioning of flywheel, piston, timing and fuel charge, the saw would likely run right up the starting rope like a 130-pound yoyo. Many of us have scars to prove it!"

By 1943 Reed-Prentice had so much war work that they decided to divest themselves of the Vancouver operation. A consortium of Vancouver businessmen purchased Smith's former Vancouver business and renamed it Industrial Engineering Limited (IEL). With the proceeds of the sale, Smith moved to Guelph, Ontario and started another chainsaw company, Hornet (see page 143).

Above and below: Model A components were made in Vancouver and the saws assembled one at a time. *Collector Marshall Trover, photos Brian Morris*

Falling with a Disston KB series. *University of British Columbia Library, Rare Books and Special Collections, images # BC 1930/292/2*

Not to be outdone by Stihl, the Disston G-10, the official saw of the US Army, boasted advanced features such as automatic rewind starter and die-cast housing. It was heavier, though.

Collector Marshall Trover, photo Brian Morris

DISSTON

Disston

In 1933 Henry Disston & Sons of Philadelphia, successful manufacturers of tools such as cross cut saws, and machinery such as chains and guide rails for chain cutoff saws that were mounted on fixed or wheeled bases, established a research and development department for the development of a portable chainsaw. The results did not set the world on fire, although they made successful models both of pneumatic and electric chainsaws.

When the war erupted in Europe, the Disston engineers studied existing machines and developed their own two-man, two-cylinder, 6-hp saws, the G-10, the G-26 with a 24-inch (60-cm) bar and the G-36 with a 36-inch (90-cm) bar, which they successfully sold to the US Army. Throughout the war and afterwards, the US Army Engineers used these Disstons at home, in Europe, and in the South Pacific.

How To Destroy Your Chainsaw

(1) *Method No. 1.* By explosives.

(a) Remove and empty the fire extinguishers.

(b) Puncture the fuel tank and fuel and oil containers.

(c) Place a 1 lb. charge of TNT between the engine and the transmission.

(d) Insert tetryl nonelectric caps with at least 5 feet of safety fuse in each charge placed.

(e) Ignite the fuses and take cover.

(f) Elapsed time should be 2 to 3 minutes if charges are prepared beforehand and carried with the equipment.

(2) *Method No. 2.* By gun fire.

(a) Remove and empty the fire extinguishers.

(b) Puncture the fuel tank and fuel and oil containers.

(c) Fire on the equipment with 50-caliber machine guns, rifles, or grenades. Fire at the engine, destroying the cylinders, fuel tank, power head, transmission, saw, guide rails, tailstock and accessories.

—Department of the Army Technical Manual TM5-4052: Saw, Chain, Portable. Section IX. DEMOLITION TO PREVENT ENEMY USE

The 6-hp Disston G-10 was produced for the US Army by the thousands starting in 1942. *Collector Marshall Trover, photo Brian Morris*

Titan of Seattle

Mill & Mine Supply of Seattle had been manufacturing logging equipment since 1905. In the mid-1930s, like so many others, MMS had experimented with its own prototype of a portable chainsaw. In 1937, they had arranged with D.J. Smith to subcontract distribution of Stihl saws to the USA. Once they were able to import Stihls, the company's salesmen, managers and mechanics, like their Canadian counterparts, began suggesting, designing and field-testing improvements.

With the outbreak of war in Europe, MMS turned its ingenuity to crafting a home-grown copy of the imported Stihl B2Z. Using a Muncie-built Neptune outboard motor engine, in 1941 they launched the 12-hp, 110-pound (50-kg) Titan Model A. By the time they had sold one hundred of the A, they were ready to launch the 9-hp Model B and the 12-hp C. All these saws were still handmade on shop benches for the professional forest industry market. Their next saw, a revised Model B (hence called "BR") incorporated a new and more reliable Tillotson "downdraft" carburetor. Even before the US entered the war and scaled up its demand for wood products, MMS had done good business with the Titans. In 1941 the firm's engineer, Max Merz, announced that the development of lighter models was being restrained only by "the scarcity of certain light metals."

Above: Stihl's Seattle distributor also jumped into sawmaking when the war started, producing Titan Model A. *Collector Marshall Trover, photo Brian Morris*

Right: The saws were heavy and so were the hats. *Collector Marshall Trover, photo Brian Morris*

Titan borrowed its design for the Model A from the Stihl B2Z, which had twin cylinders, but in the Model A's case the power was supplied by a Neptune outboard engine.

Collector Marshall Trover, photos Brian Morris

Titan 1944 Models E (top) and J (bottom). *Collector Marshall Trover,*
photo Brian Morris

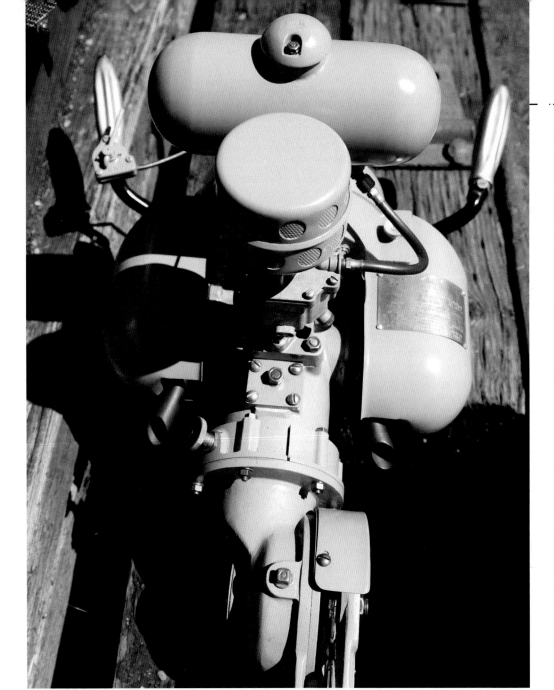

The E (darker, left and top right) was the first reliable Titan two-man model and was popular with professional fallers who needed the power of a two-cylinder motor to cut the big timber in the Northwest. *Collector Marshall Trover, photos Brian Morris*

The J (lighter) was intended as a lighter weight complement to the model E but sold in very limited numbers. *Collector Marshall Trover, photos Brian Morris*

Two-man saws proved awkward when it came to bucking the big timber of the west coast. *University of British Columbia Library, Rare Books and Special Collections. image # BC 1930/188/7*

The Quest for the One-Man Chainsaw

Although the advantages of the Stihl made it a popular model for manufacturers to copy, soon a host of variations appeared, inspired by designer ingenuity and the pressures of the wartime economy. There was a powerful incentive for saws to become smaller, lighter and more portable. There were a host of tasks both in logging and in other fields that didn't lend themselves to two-man teamwork. Pioneer logger Frank White remembers:

"Usually two fallers and a bucker was a team. That was traditional with handfalling. But when the power saws first came in, they had to have two men on the bucking. So you had an extra man. Bucking was goddamn dangerous for one guy, but now two men had to get out on the log. You know, a log lying on the level is one thing. A log lying on the sidehill all ready to roll is something else again. One guy had to be on the down-side of the log, which was no goddamn fun. Once you buck a log on a sidehill, it's gonna move, most likely. With the helper handle in the way, it wasn't easy to pull the saw out of the cut once the log started to go. They spent a lot of time chopping to free their saws with axes, or cutting them out with hand saws. They smashed a lot of them up by having logs drop on them. Smashed a lot of men up, too.

"What you'd see happening, is guys'd pull the head end off and try to run it as a one-man saw. You could rig it with a shorter bar, but one man couldn't pick it up. There was no place to put your hand. To get it up on a log by yourself you had to kind of hug it. So when the one-man saws came out, well nobody could get one fast enough."

In 1941, R.M. Burnett field-tested a one-man modification of the Shade chainsaw in the British Columbia woods. No doubt others were trying to jury-rig two-man saws so they could be operated by a single man as well, but they were too heavy and unbalanced to be very satisfactory.

This two-man Shade has had a forward handgrip added for one-man use, but it was difficult with the control hand pulling to one side and no dog-teeth for leverage. *University of British Columbia Library, Rare Books and Special Collections, image # BC 1930/391/2*

IEL (Industrial Engineering Limited)

When Reed-Prentice of Massachusetts became swamped with war work in 1942 and decided to unload D.J. Smith's old chainsaw company, by this time called Reed-Prentice BC Ltd., a buying group formed under the name of Industrial Engineering Ltd. (IEL). Principals of the new company were president George W. Sweny, vice-president George W. Thompson, secretary-treasurer A.V. (Art) Stedham, director and general manager D.J. Smith, and directors J. Lyman Trumbull and Ray A. Pitre, a mining executive fresh from the rich Zeballos gold strike. Sandy Megaw was engineer and draftsman. One of the first actions of the new management group was to open a factory service branch in Montreal in the Confederation Building at 739 Cathcart Street and stock it with machines and service parts under the management of Ralph M. Weeks. At the time of takeover two saw models were being produced: the K and the K 5000 series. The 5000 series had a welded sheet-metal fan housing and air deflector over the cylinder and a fixed-jet carburetor.

The engineering staff immediately set to work developing new models and by 1944 the 8-hp, 117-pound (53-kg) model L was introduced. A lighter saw designated model F, with a smaller engine and weighing only 85 pounds (39 kg), was also put on the market.

Meanwhile, management had been reorganized: Smith and Megaw departed and Ray Pitre took over as managing director. The company had been made an employee owned-operation with all staff having the opportunity to purchase shares, though Pitre and his wife held more than anyone else. Despite the loss of the experienced Smith and Megaw, the company was full of energy and innovation. Their next project was one that would make chainsaw history: the world's first one-man saw.

The revolutionary new machine, dubbed the "Beaver," was put through rigorous testing and sent back to the engineers for improvements. The prototype was underpowered and had no clutch, which made it dangerous to start and easy to stall when cutting. IEL increased power and added a manually operated clutch to its production model, but a more significant problem had to do with the handling of the saw.

Thus far, chainsaw engines had float-type carburetors that limited them to being operated in the upright position. This was tolerated in early two-man saws that could be set up for either falling or bucking and left in place, but it was found to be very limiting for the one-man saw that naturally lent itself to being used at different angles. Most manufacturers had solved the problem by mounting the transmission on a swivel that could

Prototype model of the 1944 IEL Beaver, the world's first one-man chainsaw. *Collector Dave Challenger, photo Mike Acres*

The circular rear handlebar was moved forward in the production model. *Collector Dave Challenger, photo Mike Acres*

The revolutionary Beaver, designed and manufactured at IEL's Vancouver factory, took the chainsaw world by storm in 1946.
Collector Mike Acres, photo Lionel Trudel

Top and above: IEL Pioneer Model AB, successor to the Beaver, with powerhead turned sideways for falling. It was the first of many IEL models to use the name "Pioneer." *Collector Mike Acres, photo Lionel Trudel*

be clamped in different positions. The Beaver had a roller-chain transmission that didn't lend itself to swivelling, so they mounted the carburetor and fuel tank on a separate unit and placed a swivelling joint between it and the engine. This allowed the whole powerhead to be swivelled while the carburetor remained vertical.

Handlebar design also came in for rethinking. The two-man saws had bicycle-type handlebars which required the operator to keep both hands on them at all times. This worked as long as there was another man to raise the bar. A one-man saw required a forward handgrip to provide the single operator with leverage for raising the bar, as well as a redesigned rear handgrip that could be operated with one hand. The Beaver prototype had a suitcase-type grip forward and a hoop-style handlebar at the rear that was very awkward to use. IEL reversed the order on its production model, putting a hoop-style bar forward so the moveable part of the saw could be gripped in any position, and a pistol-grip handle at the back. This switch worked so well that it was adopted by other makes and maintained without fundamental change for decades.

With the release of the Beaver in 1946, IEL sales quickly took off. The company opened a sales office in North Bay, Ontario to handle the pulp-cutting business in central Canada. Almost immediately, the engineering department commenced working on a new one-man saw and by 1948 the first model to carry the name "Pioneer" appeared on the market. The first Pioneer was a one-man saw that had a 3 cubic-inch (50-cc) engine and weighed 28 pounds (13 kg) without bar and chain, and it was an even better seller than the Beaver.

Despite the popularity of the one-man models, demand continued to exist for the big, heavy west coast-type saws and in 1947 IEL brought out models G and M, weighing 69 (30 kg) and 90 pounds (40 kg) respectively. Falling had always been done by two-man teams and it took some time for the habit to die out. Also, effective as the Pioneer was in eastern pulp country, there was still a lot of big first-growth timber on the coast and a 4.1-hp machine with a 36-inch (90-cm) bar wasn't enough saw to handle it. The G and M models were also used in Australia, where they were manufactured and sold as the "Bluestreak" (not to be confused with MMS/Titan's saws of the same name).

AB's 50-cc (3 cu. in.) engine would handle guide bars up to 76 cm (30 in.) and was popular in Quebec.
Collector Mike Acres, photos Lionel Trudel

INTRODUCING . . . . .

A REALLY LIGHT WEIGHT
DANARM SAW

DANARM

THE *"Whipper"*

which can take up to 30" guide blade

New sounds break the silence of the woods and forest.

The ring of the woodsman's axe now gives way to the insistent note of the internal combustion engine, and the rasp of powerful, fast-moving steel teeth sinking deep into standing timber.

This mechanical symphony can be heard in the forests of the world; in far Australia and New Zealand, in Canada, in Malayan jungle and the plantations of the British Isles.

—excerpt from the Danarm publication "Down in Three Minutes," late 1940s

In the years before World War II, T.J. & J. Daniels were a long-established foundry and engineering firm based in Stroud, a hundred miles west of London, England. They created such custom projects as a huge water pump engine (with an eight-foot flywheel) to pump water for the locomotives of the Canadian Pacific Railway. During the war their production turned to munitions and afterwards, hydraulic presses and injection moulding machines. However, at the outbreak of war in 1939 they turned their attention, as did so many other manufacturers, to producing and marketing a practical two-man chainsaw.

Cutting firewood with a one-man Danarm.

Fleeing Austria, a Mr. J. Palfi landed in London with a Dolmar Model C two-man saw. As a member of the aggressor nation, Dolmar's international copyright was void, and Palfi was hoping that the Dolmar would help get him into business in his new home. He got together with a London designer, J. Clubley Armstrong, who agreed that the Dolmar C could act as the prototype for a new, British chainsaw. Armstrong was acquainted with Lionel Daniels of T.J. & J., and soon he had reached an agreement to design a new "portable petrol driven chain sawing machine" that would be identified by a merger of their two names: Danarm.

The C, which Dolmar had produced from 1930 to 1937, featured a swivelling gearbox, a gravity-fed chain oiler, and a laminated guide bar featuring replaceable steel "wearing strips" to buffer the bar's top and bottom from the wearing action of the chain. When Danarm's first "Mark 1" chainsaw appeared in 1940, it had all these features as well, carefully copied from the Dolmar. However, where the Dolmar used an Ilo E200 engine, the Danarm used a Villiers Mark 19A.

With the escalation of the war, Danarm found an immediate market in the military. The Mark 1 and the Mark 2 (which differed only in the design of the magneto and parts of the engine cowling) were known as "XWD" saws for military use. Civilians wishing to purchase a Mark 1 or Mark 2 had to apply to the government for a licence. Another British manufacturer, Teles, built similar saws for military use by the Royal Engineers but never marketed or advertised them to the general public. Teles featured a gearbox that could be more easily removed to make the saw lighter and more portable, and also offered a more friendly arrangement for users who wanted to attach a pump or other accessory to the crankshaft.

All these models, however, had a common engine—the Villiers Mark 19A two-cycle, 250-cc (15.25 cubic inches). Bars were either 1 metre (3 feet, 3 inches) or 1.2 metres (4 feet), with a 1.45-m (4 feet, 10 inches) bar available if desired. The two-link, 3/4 inch pitch chain was 1/10 of an inch thick.

Villiers manufactured other engines besides the 250-cc model used in the Mark 1 and 2 saws. In the late 1930s it introduced a much smaller and lighter 98-cc "Junior" line of engines, culminating in the Junior-de-Luxe (JDL) model of 1940. By now, World War II was in full swing and so was the rationing of gasoline (or "petrol") throughout the United Kingdom and its allies. Villiers found its new little engine heavily in demand for motorized bicycles or "autocycles," which became vital in maintaining essential services in the beleaguered British Isles.

Transportation was not the only area to feel the need for a small, economical engine. As able-bodied males were increasingly pressed into the war effort, women moved to fill

Above and left: Like Titan and Burnett, Danarm got its start by producing a wartime copy of a German saw. The Danarm Mark 2 was closely modelled on the Dolmar Model C, and sold in great numbers to the British Army. *Collector Marshall Trover, photos Brian Morris*

Top and above: The Junior was Danarm's attempt at a lighter-weight chainsaw. *Collector Marshall Trover, photo Brian Morris*

the gaps in traditionally male-dominated trades in agriculture and forestry. Overnight, a new class of female worker was created, known as the "land girls." Competent, able-bodied and highly motivated, they were still not the heavily muscled brutes who met the challenge of the era's heavy-duty chainsaws with their own sheer bulk. To make a smaller and more portable saw, Danarm mounted a Villiers JDL engine in a tubular steel D-shaped frame with two rear road wheels. The Danarm Junior chainsaw was cooled by a fan and air cowling, its float carburetor on a swivelling intake so that it could be kept level while the saw was rotated to make horizontal, vertical or angle cuts.

With the helper's handle or "head end" mounted on the end of the bar, the Junior could make an 18-inch (46-cm) cut, but with the head end removed, it could make a 22-inch (56-cm) cut—and in doing so, became the first one-man (or one-woman) chainsaw on the European side of the Atlantic. It weighed approximately 57 pounds (26 kg).

Steel was at a premium during the war, and the grooves on Danarm guide bars were fitted with replaceable steel wearing strips copied from the Dolmar C. More remarkably, as a measure not only to save steel but to reduce weight, the Junior guide bar was steel only on the outside and in its tip, which was a steel disc. The inside of the bar was made of dense "Masonite" fibreboard. The resulting bar demanded careful maintenance. "This design worked well while the wearing strips were kept in good condition," Danarm historian B.P. Knight writes, "but if the wearing strips wore through, the inner core of hardboard was immediately damaged and the guide plate [guide bar] had to be repaired or replaced. In Australia, guide plates were usually repaired by making a new inner section of steel and re-riveting to the existing outer ones."

The Junior represented a major breakthrough when it appeared in the closing months of the war, early in 1945. By the time it appeared as a commercial model in 1946 its target audience, British women briefly allowed into "the men's trades," were gradually returning to their traditional roles.

Built near the end of WWII, the wheel-mounted Danarm Junior was meant to be used by female war-workers.

Collector Marshall Trover, photo Brian Morris

4

Emerging from War:
The Golden Age Begins

The Second World War had reshaped the chainsaw world, shifting the initiative away from shattered European economies and creating a burgeoning new industry in North America. In the immediate post-war years, patterns established during wartime continued without significant change. The companies that had sprung up to fill the vacuum left by the Europeans maintained their position by producing saws for the logging industry that looked very much like the saws going into the war—big two-man machines most useful for cutting the big timber of the west coast. Change was in the air, however, and survival would depend on how readily the existing manufacturers could adapt.

Burnett Powermatic. *Collector Mike Acres, photo Vici Johnstone*

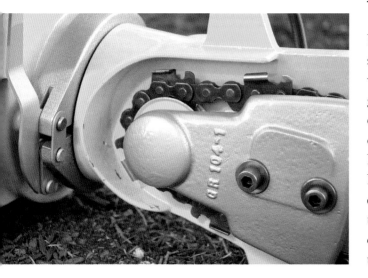

Throughout the 1950s the standards of portability and flexibility were rising, and saws became obsolete that had been state of the art a mere few years before. For years chainsaws had been started with a length of rope or a leather strap that had to be manually wound around the flywheel at each start-up attempt, then put somewhere once the saw got running. Manual starting cords were a misery, especially on the larger engines. Some engines were particularly noted for backfiring and yanking the pull-handle stingingly out of the operator's fingers or, if he hung on too tightly, climbing up the cord and wrapping his fingers around the flywheel. Plus there was the problem of the cord always getting lost, which many operators solved by slinging it around their necks. This wasn't quite as dangerous as it sounds, but it could provide quite a jolt if the operator leaned too close to the sparkplug on a wet day. In 1950 the Armstrong Company earned the eternal gratitude of all operators by introducing a reliable rewind starter in which a spring safely recoiled the cord inside a housing.

In 1951 IEL brought out the direct-drive Model HA, the first chainsaw to dispense with the bulky transmission traditionally used to boost chain torque. A cutting chain needs to be kept lubricated, and for years part of the saw operator's job was to judge when to press down the little pump handle that would squirt chain oil into the works. In 1952 Reed-Prentice introduced the Timberhog Bantam with an automatic oiler that did the job on its own. In that same year Jo-Bu introduced its Junior model to Sweden, *Chain Saw Age* magazine started up in the USA, and Shindaiwa Kogyo started making electric chainsaws in Japan. In fact, with advances in small engine technology, saw manufacturers were springing up all over the world (the USSR's Druzhba, for example, began making gasoline-powered chainsaws in 1954) because as saws became smaller, cheaper and easier to maintain and run, a huge market was opening up—not only professional loggers but farmers, landscapers and homeowners were considering the chainsaw.

Top and above: The 1949 Burnett Powermatic, a handsome machine beautifully restored by Lowell Boyd for Mike Acres.

Collector Mike Acres, photos Vici Johnstone

The 1950s and early 1960s were the golden age of chainsaw development. Activity was diversified across hundreds of manufacturers worldwide, with competitors one-upping each other with every new release. By the end of this period all the basic features of the modern chainsaw were in place.

Burnett Engineering's last saw, the Powermatic was the right thing at the wrong time—a two-man saw when the trend was to one-man. *Collector Mike Acres, photo Vici Johnstone*

DISSTON

Disston After the War

In addition to building the official saw of the World War II US Army Engineers, Disston also made saws for "the civilian market" with bars as long as 84 inches (213 cm). In 1948 the company had a banner year, introducing not only an automatic chain oiler, but the fruition of a project they had been secretly developing since the end of the war—their one-man DO-100 that with an 18-inch (46-cm) bar weighed only 37 pounds (17 kg). At the other end of the scale, in the same year they introduced "Extra-Duty Two-Man Saws" with 12-horsepower engines.

In 1951 Disston collaborated with Mercury, the well-known outboard motor company, on a new saw. The Mercury DA-211 featured Disston parts built around a 10-horsepower, 2-cylinder Mercury engine built by Kiekhaefer. Even with those impressive specs, with a 48-inch (122-cm) bar and head end on it, this two-man saw weighed only 90 pounds (41 kg), and the double-cylinder action gave its operation an unprecedented smoothness. To this, day, there are collectors who praise the DA-211 as the best two-man chainsaw ever made.

The quality of their saws held the seed of Disston's demise. They had signed a contract with Kiekhaefer that forced them to suspend production for ten years if they stopped buying the company's engines. However, the exploding market was for chainsaws that were smaller and lighter and although the Kiekhaefer engines were good, they were heavy. Stuck between a rock and a hard place, Disston abandoned the chainsaw business.

Top: Using a Disston DA-211 without tailstock to buck at O'Brien Logging, Freda Creek, BC. *University of British Columbia Library, Rare Books and Special Collections, image # BC 1930/2561*

Above: After WWII the market was flooded with war-surplus KB-series Disstons. *University of British Columbia Library, Rare Books and Special Collections, image # BC 1930/522/4234, 4235, 4238*

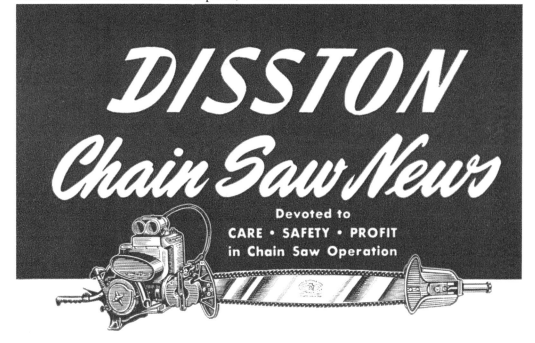

DISSTON *Chain Saw News*

Devoted to
CARE • SAFETY • PROFIT
in Chain Saw Operation

Above: The 3.5-hp Disston DO-100 (1948) and DO-101 (1950), the only one-man saws put out by Disston/Kiekhaefer, were poorly designed. *Collector Marshall Trover, photo Brian Morris*

Above: Produced in 1951 for the US Army, the DA-211was used during the Korean war. It was the last saw produced by the Disston/Kiekhaefer collaboration. *Collector Marshall Trover, photo Brian Morris*

Left: The 100 was produced in large numbers, but couldn't compete with contemporary McCullochs and IELs. *Collector Marshall Trover, photo Brian Morris*

Henry Disston & Sons had great success with this 1947 KB7AY. The engine was manufactured for Disston by the Kiekhaefer Outboard Company. *Collector Mike Acres, photo Lionel Trudel*

Don't try this when Dad's away, kids! *Photo Comox District Free Press*

TITAN

Titan After the War

By 1945 Mill & Mine Supply of Seattle were manufacturing every part of their saws except the magnetos and carburetors. With Robert Gillespie as president, Max Merz as engineer and J. Bloomer running the Titan plant, MMS created the E and the J models in 1944. In 1947, a revised Model E, the ER, was given a sexier new name and paintjob and became the Bluestreak (not to be confused with IEL's Australian model of the same name), a successful two-man saw that generated 12 hp but weighed only 60 pounds (27 kg). In 1948 they joined the boom in one-man saws with the Junior, boasting a weight of 30 pounds (14 kg) with an 18-inch (46-cm) bar. In the following year the Bluestreak Automatic featured a centrifugal clutch.

The Titan plant produced what is no doubt the longest bar in chainsaw history—a 22-foot (6.7-m) long bar operated by not just one but two Bluestreaks, one on each end. Contrived to save waste by cutting big redwoods as close to the ground as possible, Al Whyte reported that the bar "was somewhat successful but keeping the bar from side-swaying (sagging) was a problem. Once it was in the wood, it wasn't too bad." The chain was also hard to manage, liable to stretch and break.

In 1952 another Titan milestone was the Sportsman, the first one-man chainsaw made specifically for the "casual user" market: i.e. homeowners who might only use the saw a few times a year.

TITAN *Bluestreak* **CHAINS**

TITAN Bluestreak Saws are equipped with ORIGINAL TITAN CHAINS—designed for longer and better cutting and ripping. Titan chains eliminate timber bind, climbing and running . . . no swedging or setting ever needed . . . operator can sharpen chain on bar. A logger's chain.

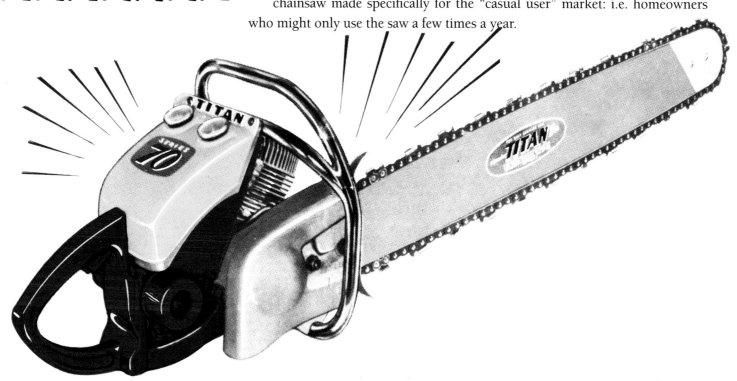

The Titan 40, a 1950 one-man model. *Collector Mike Acres, photo Vici Johnstone*

The Bluestreak was Titan's most successful two-man saw. Produced in fairly large numbers and shipped all over the US, it was very reliable and lighter than earlier models. *Collector Marshall Trover, photo Brian Morris*

The One-Man Titan

"They come out with a little one-man saw made in Seattle called a Titan. It was pretty good. Everyone got one but my brother and me. So I says to the scaler, "Fetch me out a power saw." I used it for bucking. I would fall by hand and buck with a power saw. I could buck three times as much as my brother could and he was using a crosscut saw. One day he was a long time bucking and I chopped undercuts in half a dozen trees. He still wasn't finished so I said, "God dang it, I don't know why I can't fall with that saw." I started it up and cut a tree down with it. Oh, I cut another one, I cut another one, three times faster than he could cut them by hand. So that finished it. No more falling by hand, forget it."

—Alvin Brown

If the number one problem facing any small business is failure, number two on the list is success. By the time the war was over and chainsaws were on their way from niche market to big business, MMS already faced the problem of financing a bigger manufacturing operation in order to keep up with other companies in the marketplace. To keep up, MMS needed to, among other things, switch from their established sand-casting operation to the more precise and efficient die-casting. Merz was particularly adamant about the need for this: these were still the days before direct drive, and he was excited about a new transmission he had designed—a part, however, that would have to be die-cast.

Retooling its own works was prohibitively expensive, so MMS made overtures to contract out some of their die-casting to McCulloch, a respected Los Angeles manufacturer

Above: This factory cutaway displays the internal components and rugged design of the Bluestreak. *Collector Marshall Trover, photo Brian Morris*

Below: The Bluestreak powerhead weighed 27 kg (60 lb) and delivered 10 hp. *Collector Marshall Trover, photo Brian Morris*

Titan's first one-man saw, the 1948 Junior lacked power and had overheating problems that could boil the gas in the gas tank. It evolved into the more successful models 40 and 60. *Collector Marshall Trover, photo Brian Morris*

who had done similar work for Reed-Prentice in Massachusetts. Merz made the trip to McCulloch's headquarters to try to work out a deal and was impressed by their facilities—so impressed, in fact, that he accepted their offer of a job. A version of his new transmission turned up on the first McCulloch chainsaw, the 1225A in 1948.

The loss of Max Merz and his ingenuity was not the only setback MMS had to face. Robert Gillespie was contemplating retirement and with neither of his sons especially interested in the chainsaw business, the company faced a management void. The 1953 Model Titan 100, sand-cast with two cylinders, was an anachronism in a market where lightweight portable saws with die-cast parts were setting the standard. In collector Marshall Trover's words, it "was laughed at by dealers."

Mired in debt, in the mid-1950s MMS sold the Titan line to the Draper Corporation of Hopedale, Massachusetts, which produced Penfield guide bars. In turn, in 1957 the Food Machine and Chemical Corporation of San Jose, California bought the chainsaw division from Draper (which kept producing Titan bars and chains through Penfield) for their Bolens division. But as big companies appeared promoting newer, lighter and more portable saws for homeowners and professionals alike, the end of the 1950s also saw the end of the Titan chainsaw.

Above and right: A larger 6-hp version of the Model 40, the Model 60 was only slightly heavier. Although it earned some respect among professional loggers, the heavy sand-cast design was no longer competitive in 1950. *Collector Marshall Trover, photos Brian Morris*

100

With its jazzy design aimed at the non-professional, the 1952 Sportsman was considered the first "occasional user" chainsaw.
Collector Marshall Trover, photo Brian Morris

Above: Cutaway model of the Titan 45A, a 1953 one-man model. *Collector Marshall Trover, photo Brian Morris*

The two-man Titan 100 (right and below) was an anachronism in 1953, and had flaws that caused it to be recalled.
Collector Marshall Trover, photos Brian Morris

The Titan 53 was produced in 1959 by the Propulsion Engine Corporation, who bought the Titan brand. *Collector Marshall Trover, photo Brian Morris*

Left: The last good two-man Titan. The 75 was light, powerful and reliable but had limited appeal in 1953 when the one-man saw was taking over. *Collector Marshall Trover, photo Brian Morris*

103

DANARM

Danarm entered the one-man chainsaw derby with its 1949 Type C, marketed as the "Tornado."

Danarm After the War

Although one-man saws were appearing here and there, the chainsaw was still seen—perhaps because its predecessor, the two-man cross cut saw, was still very much in use—as essentially a two-person machine. One-man saws were luxuries, sportscars in a market ruled by the sober and substantial family station wagon. They were also still heavy enough to allow two-man saws to remain a viable alternative. After the war, Danarm's next saw was the Mark 3, which was no lightweight. It was available in a range of sizes, weighing in at 118 pounds (55 kg) with a 30-inch (75-cm) bar and a Villiers Mark 25A 250-cc engine, and gradually getting heavier until it reached 166 pounds (75 kg) with an 84-inch (215-cm) bar and a Mark 27B 350-cc engine.

In using Villiers engines, which weren't following the trend of casting their engine parts with the new lightweight alloys, Danarm's chainsaws were staying good and heavy while all over the world the saws were getting lighter. As a result, the Mark 3 must have been a real burden for its hard-working users, no matter how big and strong they might be. But Danarm historian B.P. Knight praises the machines: "They were of relatively simple construction, had a reliable engine, were easily maintained by timbergetters and generally gave years of trouble free service." *DOWN! In Three Minutes* shouted the banner headlines on Danarm ads, boasting that three minutes was all it would take for the Mark 3 to fell a tree one yard in diameter.

Danarm's success was by no means lost on Villiers. Although primarily intending its engines to be used in motorcycles and the smaller "autocycles," Villiers had found that the use of its engines in chainsaws was proving to be a market of no small significance. After the war, when Villiers developed its new Mark 2F autocycle engine, it designed a variation, the Mark 3F, exclusively for chainsaws.

Using the 3F, in 1949 Danarm premiered its Type C under the name "Tornado." This saw was far more powerful than the Junior, almost 4 hp, and weighed less—with a 22-inch (56-cm) bar, only 49 pounds (22 kg). Although the Canadian models had not been successful, they had offered features that Danarm sensibly incorporated into the Tornado. Like the Danarm 98, the Tornado had a large looped front handlebar that doubled as a chain oil reservoir. The carburetor and fuel tank could be similarly swivelled and locked into any position for cutting at any angle. For a change, instead of a heavy cast-iron cylinder, the Villiers Mark 3F had a light alloy cylinder with—to reduce friction—chrome-plated walls (the result referred to as a "honey-chrome bore").

The advertisements boasted "The World's Fastest One-Man Chain Saw" or simply "The Fastest Saw in the World!" It was available with a "helper's handle" or head end to be attached to the end of the guide bar for two-man operation, an optional stub handle fitted at a right angle to the pistol grip and "in accordance with the latest Canadian practice," a four-toothed "dog" to help pivot the saw while cutting.

The Tornado was a successful model, but in the face of mounting international competition from other companies forwarding their own one-man chainsaws, Danarm discontinued it in 1955. This may have made sense in terms of its market in the United Kingdom and Europe, but it annoyed a lot of people Down Under. The Tornado had been an immediate success when the first saws had hit Australia in 1950. Danarm pulled it just as sales were taking off.

Because of the weight of internal combustion engines, electric saws had a special appeal in the early days, as long as the problem of an external power source could be solved. Back in the UK, Danarm offered electric versions of the Mark 1, Mark 3 and Junior saws. In the mid-fifties it came up with a "Baby Electric." In the early 1950s, Danarm also introduced its "Pneumatic" model, available with guide bars ranging from 16 to 23 inches (46-58 cm). This saw, of course, needed to be connected to a heavy-duty compressor. Chain oil was stored in the tubular front handle, and fed to the chain via pressurized air. Like previous pneumatic models, this one could be run underwater.

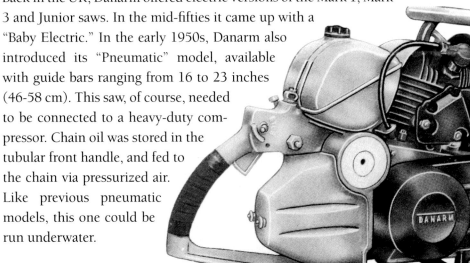

Danarm was an early producer of one-man saws.

Danarm in Canada

Throughout the 1940s, Vancouver, BC was an important centre of world chainsaw action. World War II, with its uptake of manpower and its demand for lumber, together with its interruption of German imports, had created the right conditions for a homegrown industry. As it turned out, there was sufficient interest and experience among local entrepreneurs such as R.W. Shade, D.J. Smith and R.M. Burnett to take full advantage of this opportunity.

After the war, the pace barely slackened. IEL and PM, the two leading manufacturers, built on their wartime momentum with a bold program of innovation and expansion, becoming world-scale players in the process. Danarm decided to see if the success it had enjoyed with its British saws could be repeated in this vortex of chainsaw-making. In about 1947 the Danarm name first appeared on a one-man saw engineered by The Hi-Baller Co. Ltd. of Vancouver. The Danarm 98, like the British "Junior" model, used a Villiers JDL engine and could take up to a 30-inch (75-cm) bar. However, its heavy cast-iron cylinder, fine for powering a bicycle, was too slow and cumbersome to make a competitive chainsaw. Like the smaller version of Britain's Mark 3, the two-man Danarm 250 used a Villiers Mark 25A 250-cc engine and could run a 48-inch (120-cm) bar with 3/4-inch (19-mm) pitch chain.

These saws benefited from being at the centre of the chainsaw world, but they were also surrounded by tough competition, and eventually, Danarm's efforts proved to be no more than a footnote in the history of North American chainsaws. At home, however, and in Australia, they were to be a viable entity for decades to come.

Chainsaws in Australia

Early chainsaws introduced by Rinco, Stihl, Dolmar, Wolf and others never made it to Australia. In the days before commercial air travel, Down Under was indeed far removed from the manufacturing centres of Europe and North America—a market not only too distant for immediate exploitation but underpopulated to boot. Despite his bottomless capacity for enterprise, Andreas Stihl never made it to Australia during his many sales trips during the 1930s and once war broke out in 1939, it seemed he never would.

However, the success that Danarm enjoyed in England soon prompted it to search for overseas markets: Australia with its temperate and tropical forests (and, as a British Com-

106

monwealth member, minimal import tariffs) presented a bounty of opportunity. In 1942 Veneer and Woodworkers Supply (VWS) of Sydney imported a shipment of Danarms—the first chainsaws to reach Australia. Business was good throughout the war years.

Then in 1949, with the death of VWS's original owner, the business was purchased by an E.J. Simmer, a French-born engineer who emigrated to Australia in 1928, became a citizen and later, the first person of non-British origin to attain a military commission, when he became a squadron leader with the Royal Australian Air Force. Simmer was also a canny businessman, and when the popular Tornado model was discontinued, he was quick to take the initiative. He set up a factory in Camperdown, a suburb of Sydney, ordered a shipment of Villiers 3F engines from England, and began manufacturing his own Tornados.

The domestically grown saw prospered from Simmer's acumen. Although the Tornado was already relatively light and powerful, in one advertisement Danarm used a very large man to pose as the operator in order to underplay the machine's actual size. Danarm's Down Under operations grew during the 1950s and in 1962 it introduced its first domestically designed saw, the DV700.

Above and left: Canadian version of the Danarm 250, marketed in Vancouver by the Hi-Baller Co., differed from the English version only in gas tank and tailstock. *Collector Marshall Trover, photos Brian Morris*

The 1947 98 Hi-Baller was made in England by Danarm and supplied to the Vancouver-based Hi-Baller company. A viable one-man saw for the time, it used a motor scooter transmission and clutch. *Collector Mike Acres, photo Lionel Trudel*

Chipper Chains

We tend to build new technologies on old ones, and so it was with the design of saw chain. Mike Acres writes:

"Naturally the chains were copied from various cutter types used on hand saws of the day. The 'scratch' or 'standard' chain was a series of cutters and rakers as found on a crosscut saw. These chains cut quite well at the speed they were being driven which could be as much as 800 FPM (feet per minute)."

The scratch chain had limitations, however.

"For one thing, they had a lot of teeth, so it took a lot of filing to maintain them. If you hit a rock while bucking, it could take considerable downtime before you were cutting again. Filing itself was (and is) a specialized skill."

Scratch chains were also at the mercy of the grain of the wood they were cutting. They could only cut efficiently at a ninety-degree angle to the grain. In contrast, the curved cutters on modern chains can saw from any angle. Their secret lies in a cutting edge which is curved or bent over in an 'L' shape so that it strikes the wood horizontally. Unlike the scratch chain, which used vertical, knife-like teeth to slice into the wood, each tooth in a "chipper" chain bites a small chip out of the wood.

The chipper design is usually attributed to Joe Cox, a Portland logger who had been tinkering with new chain designs for months when, according to legend, he stopped to take a close look at a timber-beetle larva that was boring through a nearby tree stump. The grub's opposing C-shaped mandibles seemed to be doing such a good job that Joe set about creating a mechanical equivalent in his basement workshop. A year later, he was successfully selling the Cox Chipper Chain.

Although this is an appealing story with its "Eureka!" moment, it is best appreciated while bearing in mind that the previous decade had seen prototypes of designs similar to Joe Cox's chipper chain. There is evidence to name the J.E. Hassler company as pioneer of an L-shaped cutter design in 1940.

Regardless of the sources of his inspiration, Joe turned out to be a good businessman. In 1948 he moved his newly founded Oregon Saw Chain Manufacturing Corporation out of his basement and by 1950 he had seventy employees making and selling chain, guide bars, sprockets and other accessories out of a twenty-two-thousand-square-foot factory. In 1952 Oregon bought out the Planer Chain Ltd. of Guelph, Ontario, and in 1957 the company changed its name to Omark Industries, all the while keeping the Oregon brand name on its products.

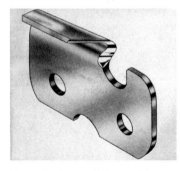

Above: Cutting tooth from a chipper-type chain showing curved profile.

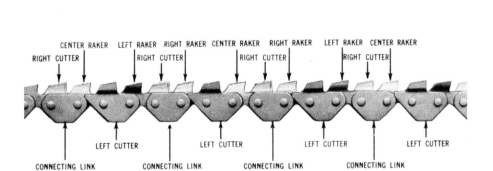

Above: Chipper chains were easier to guide and sharpen.
Left: Disston straddle chain with straight cutters and rakers, typical of standard or scratch chain in use until mid-1940s.

The Mall Tool Co. had been making high-quality tools for 20 years before producing their first chainsaw.

Later in life, Arthur Mall became good friends with Andreas Stihl—a relationship that worked to their mutual benefit. Mall had been born in 1895 in Hammond, Indiana, but had lived in Chicago after the age of two. As a six-year-old boy he sold newspapers on the streets of the southeast side, and after high school he attended Steven's Institute in Hoboken and Armour Institute in Chicago before graduating from the navy's Officer Candidate School. He served in the navy from 1916 to 1918, and within a few years was in business for himself.

On January 1, 1921—years before Stihl built his first saw—Mall established the Mall Tool Company in Milwaukee, Wisconsin. Mall's specialty was a flexible driveshaft that, first applied to a sanding and polishing tool, brought him a brisk business in the booming Detroit auto industry. Within a year Mall had moved his business to Chicago and started to adapt his flexible-shaft design for electric, pneumatic and gas-powered tools.

Like many self-made successful businessmen, Mall was a hands-on sort. While his competitors tended to offload supplies and service to local distributors, Mall opened his own factory service outlets in major US and Canadian centres, and tended them regularly in his own company airplane. Soon, Mall opened a second factory in Toronto in order to better deal with the Canadian tool market. During the 1930s Mall evidently adapted one of his flexible driveshafts to run a chainsaw cutting attachment, but otherwise seemed to be in no hurry to launch himself into the chainsaw business. For one thing, while the trend in saws was moving toward two-stroke gasoline engines, Mall products were generally driven by electric motors, with Briggs & Stratton four-stroke engines on their gas-driven units, such as concrete finishing trowels.

In 1939, just before the USA closed its ports to German products, Mall bought five Stihl saws. However, it wasn't until 1941 that Mall found the time to produce his own gasoline engine and install it in the Mall Model 4 chainsaw, and it wasn't until almost the end of the war that Mall really got into the saw business, introducing the electric Universal chainsaw that sold well in the construction market. He also refined his Model 4 to create the 5. Although a two-man saw, with its chrome-plated aluminum cylinder and magnesium castings, the 5's powerhead weighed a mere 52 pounds (24 kg). It was offered with 18-inch (45-cm) to 72-inch (182-cm) bars, using its own Mall scratcher chain. The model 5 was manufactured in an army version that was purchased by the US Army later in the war.

After the war Mall wholeheartedly welcomed back returning employees, giving them

each packages of farmland acreage outside of Chicago in Crete (now Park Forest) that he had purchased during the war. However, not one of the employees wanted to start a family farm. They opted for cash settlements, and Mall turned the acreage into experimental farms, raising grain, hogs, beef and dairy cattle and testing saws and electric hand tools.

Equally wholeheartedly, after the war Mall seriously entered the chainsaw business. "The demand in gasoline chainsaws," Mike Acres writes, "was for lighter machines," and Mall's quarter-century in the tool-making business had given him long expertise in sand-casting and a considerable head start in die-casting and working with the new lightweight metals and alloys. In 1945 came the Model 6; a year later the Model 7, with a die-cast magnesium fuel tank among other improvements, was offered with a bar as long as 12 feet (144 inch/365 cm). The quality of the electric Universal was continued with the electric 4E and in 1951, the 5E. Meanwhile the Mall catalogue was filling up with gas earth augers, pumps, concrete vibrators and "pinchless" chainsaws in the bow saw design. In 1950 the gas chainsaw 7G featured, instead of a starter rope that had to be manually wound each time, an automatic rewind starter. Innovations such as these made Mall a popular model for the lumber industry. The company also produced its own line of bars and chains.

Mall's last two-man saw was the Model 8, produced in 1955. By this time, the company was already established in the one-man saw business. In 1949, its horizontal-cylinder Model 10 had an optional head end that could be removed for one-man operation, although with a 24-inch (61-cm) bar and chain the saw weighed a challenging 58 pounds (26 kg). Mall concentrated on this option; even its 1954 OMG saw, which weighed a more manageable 34 pounds (15.4 kg) with an 18-inch (45-cm) bar, maintained the slot in its bar tip for fitting a helper handle, just in case.

By 1951 the one-man saw boom was on and Mall bought out D.J. Smith's Guelph, Ontario firm Hornet, which since 1948 had been doing well with its DJ3500, and merged its operations with his Toronto factory. In 1955 Mall produced the 105-cc (6.5 cubic inches) 2MG engine, which besides a chainsaw attachment good for a 36-inch (90-cm) guide bar, was capable of powering a two-man earth auger, a wood drill transmission and a flexible-shaft sump pump.

However, in 1956, after thirty-five years in business, Mall sold out to the arms manufacturer Remington. For several years saws were manufactured under the dual Mall/Remington brand name. Then the Mall logo gradually disappeared except on such products as the still-popular 2MG engine, which was used until 1969 to power the portable diamond drill manufactured by the Minogue company of North Bay, Ontario.

Above and below: Mall's Universal electric chainsaws were durable and popular. Arthur Mall's experience in general tool-making served him well in the competitive chainsaw business. *Collector Mike Acres, photos Vici Johnstone*

Above: A chrome-plated aluminum cylinder and magnesium castings made Mall's second gas-powered saw, the 1944 Model 5, relatively light at 52 pounds (24 kg). It was built for the US Army during WWII. *Collector Marshall Trover, photo Brian Morris*

Right, far right and below: The 1945 model 6 was an improvement over the model 5 and earned Mall a good share of the post-war market for two-man chainsaws.

Collector Marshall Trover, photo Brian Morris

Left: Mall Model 10, here with bow attachment, was a lighter saw for one- or two-man use.
Collector Marshall Trover, photos Brian Morris

Below: The Mall 7H was a 1947 update of the highly successful Model 7 with an automatic rewind starter.
Collector Marshall Trover, photo Brian Morris

The Model 12, Mall's first one-man saw.
Collector Mike Acres, photo Lionel Trudel

Above: the 1955 3MG. *Collector Marshall Trover, photo Brian Morris*

Right, from top: 1MG, OMG, 2MG. *Collector Marshall Trover, photo Brian Morris*

Below: the 4MG, Mall's first direct-drive model. *Collector Marshall Trover, photo Brian Morris*

Left, top: A nicely restored 1953 Mall 2MG.
Collector Mike Acres, photo Lionel Trudel

Above and bottom left: 1955 Mall GP, a 26-lb. (12-kg) direct-drive saw with all-position diaphragm carburetor. *Collector Mike Acres, photos Vici Johnstone*

McCulloch saws raised the bar for technological innovation.
University of British Columbia Library, Rare Books and Special Collections, image # BC 1930/135/3

Legendary west coast boss logger P.B. Anderson checks out a new McCulloch at a Victoria trade show. *University of British Columbia Library, Rare Books and Special Collections, image # BC 1930/3/83*

McCulloch

In the 1930s, championship hydroplane racer Robert Paxton McCulloch set up a machine shop in Milwaukee, Wisconsin to manufacture engines for midget racers. Despite the small scale of his enterprise—when the lathe was being used, anyone entering McCulloch's office had to duck under a turning shaft—the business prospered because of its owner's talent for designing and manufacturing small engines. In 1943, he sold his first business to Borg-Warner, and six months later founded McCulloch Aviation Inc., supplying the Los Angeles company Radioplane with generators as well as engines for some six thousand target drones used to train American fighter pilots. Their success was such that Robert McCulloch was convinced that the company's future lay in lightweight engines and moved to a new 80,000-square-foot workspace in Los Angeles. After the war Reed-Prentice contracted McCulloch to produce engines for a chainsaw, and by 1947 McCulloch was finishing the designs for its own line of saws. In 1948, when his contract with Reed-Prentice expired, McCulloch put the lessons he had learned to good use.

In July the Los Angeles plant produced the McCulloch 1225A. With its automatic clutch and swivelling gearbox (its design courtesy of Max Merz, recently arrived from Mill & Mine Supply in Seattle), it was still a two-man saw, but a two-man saw that weighed less than 50 pounds (22.7 kg). The 1225A designation was soon changed to 5-49: 5 horsepower, 49 pounds (22.2 kg). Setting the style for McCullochs to follow, the saw was painted a bright enamel yellow. The 5-49 was followed by the 7-55 two years later with a larger engine and a built-in chain oiler. McCulloch continued to build two-man saws with the 99, the 1-92, the 1-93 and finally the 940, which continued until 1971.

Dave Challenger

"He (McCulloch) was very innovative. He came in with die-cast castings when everyone else was using sand-cast castings. That makes a big difference in weight. Beavertail bars was another of his ideas. He made his own carburetors. Everybody had been using Tillotsons prior to that. Quite a few little things. They were the king of the heap when I started in 1951. They were hard to work on, Jesus. The bolts going into the starter housings, they had captive nuts behind them. If you ever dropped that out of there, of course you couldn't tighten the goddamn screw. It was a nightmare. The carburetors weren't as good as they should have been and the ignitions got damp. They were always a problem. But they developed things other guys just hadn't done at all."

—Dave Challenger, ex-dealer, collector

The next goal was a practical one-man saw. This meant, first of all, reducing weight, something McCulloch was well positioned to do. In the aviation business where weight was even more critical, McCulloch had become adept at die-casting aluminum and magnesium, and he made full use of these skills in his revolutionary Model 3-25 that, even mounted with an 18-inch (46-cm) or 24-inch (61-cm) bar and chain, still weighed under 25 pounds (11.3 kg).

Another key innovation McCulloch brought to the chainsaw was the diaphragm carburetor. In the simplest design, a design essentially universal up until about 1950, fuel was fed from the gas tank into the carburetor's float bowl. In the bowl, a float on the surface of the gasoline served to regulate its passage through the carburetor and into the cylinder. It is a simple and efficient system, but depends on gravity, so it only works when the carburetor is in an upright position.

Therefore, in contrast to modern saws that will run at any angle, early chainsaw engines had to be kept upright. To move from falling to bucking, the cutting attachment had to be rotated using a swivelling mechanism usually placed between the powerhead and the transmission. However, the engineers at McCulloch had also been working on this problem, and the carburetor on McCulloch's 1949 flagship chainsaw, the 3-25, replaced the float system with a flexible diaphragm to regulate the gas flow. Since this action was unaffected by gravity, 3-25s and subsequent McCulloch models could make angled cuts with no need to swivel the bar, eliminating another heavy mechanism and simplifying use. In time Tillotson, the main carburetor supplier to the chainsaw industry, developed an improved diaphragm carburetor with a built-in fuel pump, and the diaphragm carburetor became an industry standard.

A saw that could be tilted on its side needed a different handlebar configuration than former straight-up types, and the 3-25 introduced front handlebars on the sides as well as the top so the machine could be comfortably gripped at all angles. The rear handgrip was of the single pistol-grip design, placed low so the machine could be rotated without changing the centre of gravity. Like most of the McCulloch innovations, this one soon became standard throughout the industry.

The chainsaw was beginning to settle into its modern configuration. With the introduction of the 3-25, McCulloch became a pivotal force in the

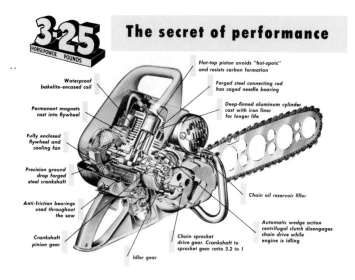

Drawings of two cutting-edge McCulloch saws, the 3-25 (above) and the 1225A (below.)

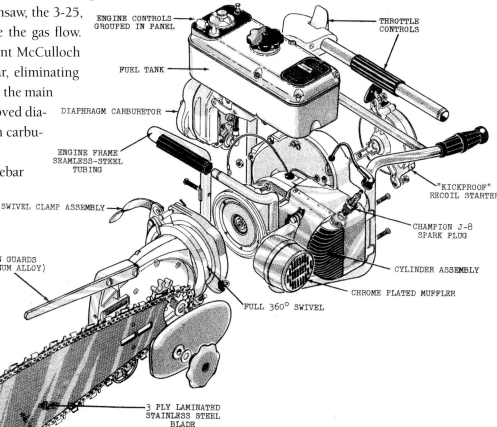

McCulloch reshaped the chainsaw world when they brought out the ground-breaking 1225A in 1948. It was light, powerful and packed with innovations. *Collecton Marshall Trover, photo Brian Morris*

Top and above: McCulloch's 1225A had die-cast magnesium housings for lightness and an all-position diaphragm carb for versatility. Its handlebar design allowed for one- or two-man operation. *Collector Marshall Trover, photos Brian Morris*

use of chainsaws in all areas of the lumber industry. The 3-25 was followed by the 47 in 1953, which was a very successful model through 1956. The 4-30 (later the 4-30A) came out with a slightly larger engine in 1953. In 1955, McCulloch brought out a one-man saw for use in big timber, the 7.3-cubic-inch (120-cc) 73. Meanwhile McCulloch didn't neglect the small-saw market and in 1952 brought out the 33, a nice little 3.3-cubic-inch (54-cc) gear-drive. The 33A, 33B, Super 33, D33, 35, 39 and Thrifty Mac all followed before the series was discontinued in 1959.

In 1956 the company came out with a hotter engine using innovative reed valves with a third-port transfer induction system and had an instant success in the 19-pound (8.6-kg) D44. This design was carried through all of the professional saws built by the company until 1977.

McCulloch introduced possibly the most revolutionary chainsaw engine ever with the 1962 BP-1. BP stands for "balanced piston." Normally a one-cylinder engine has a counterweight on the crankshaft to offset the weight of the piston, but it is not perfect. In the 2.7-cubic-inch (44-cc) BP-1 McCulloch added a second piston of the exact same weight as the first piston. The balance was so perfect the BP-1 would instantly rev to 10,000 rpm; in fact it would rev so high that if the governor failed to kick in it would explode, which caused it to be discontinued.

In 1965 McCulloch re-introduced a 3.3-cubic-inch (54-cc) series, the 1-10, 2-10, 3-10, the electric-start 3-10E, 4-10, 5-10 and 5-10E (electric start). The 6-10 had a 4.3-cubic-inch (71-cc) engine and was followed with the 7-10 series. The 10-10 series started in 1971 and continued through the PM-10-10S in 1998. Also in this series were the CP55, PM55, PM60, CP70, G70, SP70, SP80, PM555, PM570, PM700, PM800 and PM850S, which was built until 1994. In the field of consumer saws, McCulloch built a long list in the "Mac" series, the most famous of which was the Mini Mac 6, thanks to a prize-winning TV ad starring two beavers who wished to get one for their toothless grandpa.

There is much more to the McCulloch story. A swashbuckling entrepreneur in the Howard Hughes mould, Robert Paxton McCulloch made a fortune supplying turbochargers to Detroit and spent much of it trying unsuccessfully to break into outboard motors, buying Scott-Atwater. He developed real estate in Arizona and was responsible for moving the London Bridge, brick-by-brick, from London to Lake Havasu. He sold his company to Black and Decker for $65 million in 1973 and died in 1977. Black and Decker sold the firm to a consortium of employees, who were in turn forced to sell it to Shop Vac Corp. In 1998 the one-time king of chainsaw makers was closed by the banks and the rights to use the name in the US were auctioned to the Jenn Feng Company of Taiwan. Electrolux picked up rights to use the name outside the US.

McCulloch's first one-man chainsaw was as revolutionary as their first two-man. With 3 hp, only 25 lbs. (11 kg) in weight and the ability to run in any position without swivelling the bar or carb, the 3-25 smoked the competition and put McCulloch on top of the chainsaw world, a position they held for two decades. *Collector Marshall Trover, photos Brian Morris*

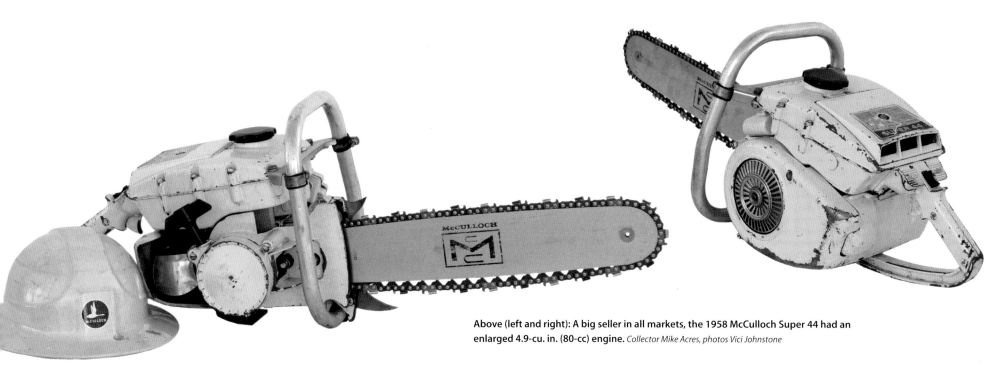

Above (left and right): A big seller in all markets, the 1958 McCulloch Super 44 had an enlarged 4.9-cu. in. (80-cc) engine. *Collector Mike Acres, photos Vici Johnstone*

Above and left: The 1952 McCulloch Super 33, the first version of the lightweight model that had a twenty-year model-run to become one of McCulloch's most popular series.

Collector Marshall Trover, photos Brian Morris

122

Left and right: The 1977 Super Pro 125C was a 123 cc model for pros. *Collector Lowell Boyd, photos Vici Johnstone*

Right and below: One of McCulloch's most brilliant and least successful saws, the 1962 BP-1 had uniquely balanced cylinders that eliminated vibration but risked runaway, causing it to be withdrawn. *Collector Marshall Trover, photos Brian Morris*

A very fine mid-size saw, the 10-10s was first built in 1967 with a 3.3-cu. in. (54-cc) engine. The PM10-10S with a 3.5-cu. in. (57-cc) engine continued in production until mid-90's. *Collector Mike Acres, photos Vici Johnstone*

IEL After the War

Even after introducing its trend-setting one-man Beaver and Pioneer saws immediately following the war, the innovative Vancouver manufacturer IEL continued to make heavy two-man saws for several years. The writing was on the wall for the heavy brutes, however, and in 1949 IEL produced a saw that split the difference between the smaller one-man saw and the old two-man dinosaurs. This was the Pioneer Twin, a 45-pound (20-kg), 6 cubic-inch (98-cc) model capable of handling a 60-inch (1.5 m) bar, and it could be operated by either one man or two. The first one, which used two Pioneer AB-sized cylinders, was a bit underpowered so it was quickly followed by a larger-bore model called the Super Twin, with 9.82 cubic inches (161 cc) and available in either the one-man handlebar version or two-man handlebar version. Beaver-tail cutter bars up to 60 inches (1.5 m) were offered for the one-man version, and up to 84 inches (2.1 m) with tail stock for the two-man version.

In 1951 IEL introduced a new one-man saw built around the same larger 4.91-cubic-inch (80.5-cc) cylinder, the Super Pioneer. With rewind starters, sleek, unitized designs, and the fire-engine-red enamel finish that would become the Pioneer emblem, both one- and two-cylinder models looked like the saws of a new generation. The success of this new line, and the new name, put IEL into high gear, making it one of the world's most successful chainsaw manufacturers.

Direct Drive

The early saws had, in Mike Acres' words, "just enough power to pull the chain of the day through a five-foot cut if it ran well and the chain was sharp." In the days of the two-man saw the bar had to be longer than the tree was thick. This meant that in the course of falling a tree five feet thick there would be up to five feet of chain in contact with the wood. In addition, the teeth themselves were extra large. Chain size is measured by "pitch," the length of a link from rivet to rivet. The first chains had 3/4-inch (19-mm) pitch, twice what a typical saw might have today. This created so much drag the early saws had to have big 250-cc motorcycle engines, but even at that they had to be geared down to give them enough torque (rotary force) to keep their long, heavy chains cutting. For this reason all chainsaws up to 1951 had transmissions using a reduction system of gears, belts or chains to increase torque. These transmissions added weight and imposed another bulky component between the saw and the work.

In the early 1950s, engineers Frank Davison, Cecil Cookson and Hayo Deelman at IEL had a brainwave. If they reduced the size of cutting teeth as well as bar length, drag would reduce

Close ties to the west coast logging industry allowed IEL to tailor their saws to loggers' needs. *University of British Columbia Library, Rare Books and Special Collections, image # BC 1930/522/4234, 4235, 4238*

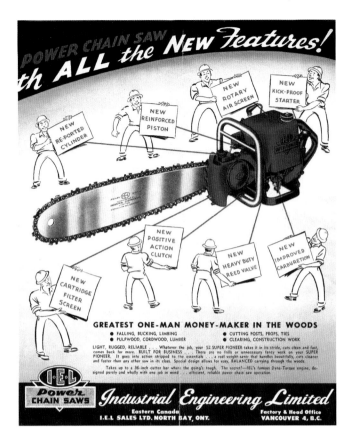

Bucking with an IEL Super Twin.

sufficiently enough that the saws would not need transmissions. In 1951, IEL introduced the Model HA, the world's first direct-drive chainsaw. It had an 88-cc motor, .400-inch (10-mm) chain and would handle up to a 32-inch (81-cm) bar. The smaller teeth caused some scepticism, but the HA chain ran so much faster than geared saws that it set a new standard for cutting speed. In time most other firms followed IEL's lead and eliminated the transmissions on their one-man saws, and chain size continued to go down while chain speed went up. Designers also learned how to get more torque and higher speed from smaller, lighter engines. Today a five-foot cut could be made relatively easily with, say, an 80-cc saw with a chain of 3/8-inch (8.5 mm) pitch on a 36-inch (90-cm) bar while a professional faller cutting a lot of big timber might use chain as large as .404 inches (10.3 mm)—and a direct-drive saw with a 99-cc engine swinging a 48-inch (122-cm) bar.

In 1951 the company scored another major breakthrough with the world's first direct-drive saw, the HA. This high-revving, fast-cutting one-man saw was especially effective for pulp cutting. The light one-man chainsaw had taken over much of the hand work in cutting pulp, but there was still one major task that had to be done with an axe—limbing. The slow chain speed of the gear-drive saws made it hard for them to penetrate the hard wood of limbs. They tended to bounce and snag. Limbing with chainsaws was outlawed in some jurisdictions for safety reasons, but it was soon discovered that the HA, because of its high chain speed, could cut through limbs efficiently and safely. This revolutionized pulp cutting—once IEL was able to have the safety regulations changed. Eventually direct drive became standard on all saws, though it was some years before manufacturers were able to achieve the right balance of torque and engine size to introduce it in the bigger saws used in the west coast logging industry.

The pace of development did not slacken at IEL. The HA direct drive was replaced by the HB, which featured the newly developed Armstrong rewind starter. To this point, all IEL rewind starters had used a steel cable. The Armstrong starter used a nylon cord that was far more flexible and much easier to replace if broken.

The next big change to come about in chainsaw development was the all-position diaphragm carburetor. Originally developed for airplanes and introduced to chainsaws by McCulloch, diaphragm carburetors became available to the rest of the industry when Tillotson introduced its H series in the mid-fifties. The first IEL model to use a Tillotson diaphragm carburetor was the 1955 model HM, its first direct-drive, all-position saw. Tillotson made a significant improvement to its diaphragm carburetor by adding a built-in fuel pump, and in 1956 IEL released an upgrade of the HM fitted with an improved carburetor, the HC. This saw combined the two most important innovations of the day and was such a hit the IEL factory had to put on three shifts to try to keep up with demand.

The model lineup of gear-drive models continued with an increase in cylinder bore from 2 inches (50.8 mm) to 2.065 inches (52 mm). Thus the Super Pioneer DB was replaced with the Super Pioneer DC, and the Super Twin with the Super Twin PG. This same engine was marketed with a portable fire pump that was used in the British Columbia woods for years, long after the Super Twin chainsaw itself had passed out of use. The last two gear-drive models in the IEL lineup were the DD and the PH, which had higher gear ratios and Armstrong starters.

Up to this stage the market for larger logging saws was still dominated by gear-driven models, as no manufacturer had produced a large direct-drive saw. IEL, which had been out front in the shift to direct drive, maintained their lead with the 1956 release of their next breakthrough model, the 7.68-cubic-inch (126-cc) direct-drive JA, capable of operating with a 50-inch (127-cm) bar. IEL's new models for 1957, the RA and JB, were advertised as "The World's Best Chain Saws," and it was a legitimate claim.

IEL kept loggers happy by thoroughly field-testing new models like this Super Twin. *Steelworkers Local 1-80*

With these models, IEL had essentially created the modern chainsaw. In the decades since, most of the developments—higher performance engines, lighter castings, electronic ignition, anti-vibe mounts, chain brakes, etc.—have been more in the order of refinements than fundamental design changes. It was an impressive accomplishment for a small, employee-owned manufacturer far from the centres of world commerce and a great credit to the engineering team of Frank Davison, Cecil Cookson and Hayo Deelman.

Equally impressive was the way they had done it. In the mid-fifties IEL opened a new factory in the Vancouver suburb of Burnaby where it made all its own castings in its own foundry and even made its own guide bars and chain. IEL seemed the very soul of independent enterprise.

Then in mid-1956 everything changed. The giant US manufacturer of Evinrude and Johnson outboard motors, Outboard Marine Corporation (OMC), made an attractive offer of $23.75 per share to purchase the feisty little Canadian manufacturer. This was worth over $2.7 million to the Pitres, a sum that would be hard to ignore in 1956. Even a secretary, Lois Adams, had 5,200 shares worth $123,500—enough to retire on at that time. A vote was put to the shareholders and when it was counted the decision was to sell. This was effectively the end of IEL as a Vancouver-based company. The new owners changed its name to Pioneer Saws Ltd. and in 1958 moved operations to the existing OMC factory in Peterborough, Ontario. There the Pioneer name continued to outshine its competitors as the leading Canadian chainsaw manufacturer for another twenty years.

With new direct-drive saws like the 1957 IEL RA Pioneer, one man could do the work three had done a few years earlier—and do it faster. *University of British Columbia, Rare Books and Special Collections, image # BC 1930/440/9534*

From a tiny, employee-owned startup in 1943, IEL had grown into a global chainsaw force by 1956—then the big boys came calling.

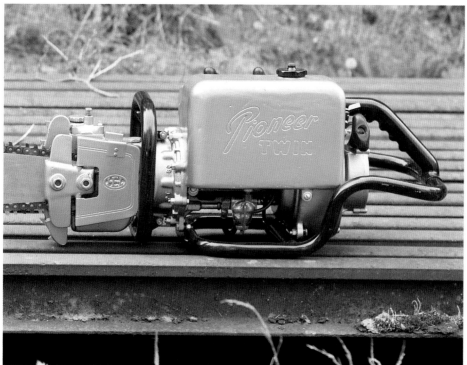

Left and below: In 1949 IEL brought out their first in-line twin cylinder model, the Pioneer Twin. It was underpowered but served as a test bed for many new ideas. At 20 kg (45 lb) it could be used by either one or two men. *Collector Marshall Trover, photos Brian Morris*

With the 1951 Super Twin, IEL addressed the deficiencies of the Pioneer Twin, creating a powerful and reliable saw that became a staple in the forest industry. *Collector Mike Acres, photo Lionel Trudel*

Above and top: The Super Pioneer, first introduced in 1951, won IEL many fans as a professional falling and bucking saw noted for its durability. The 80.5-cc engine was used to power other devices such as boom stick augers and diamond drills. *Collector Mike Acres, photos Lionel Trudel*

Right: Model HA. *Collector Mike Acres, photo Lionel Trudel*

In 1951 IEL led the world into the direct-drive era with its model HA, the world's first direct-drive chainsaw. For falling, the powerhead and bar swivelled sideways, leaving the tank and carb upright. *Collector Mike Acres, photo Lionel Trudel*

Top, above and left: The 1955 Pioneer HM combined the two major innovations of the day: direct drive and the diaphragm carburetor, permitting it to cut in any position with no swivelling. Especially popular in small tree areas. *Collector Mike Acres, photos Vici Johnstone*

Above and above right: A powerful 100-cc mid-sized direct drive, the high-performance Pioneer RA made IEL one of the leading chainsaw manufacturers in the world. *Collector Mike Acres, photos Vici Johnstone*

Above and left: In 1956 the IEL factory had to work three shifts a day to meet the demand for this rugged and reliable direct-drive Model HC.

Collector Mike Acres, photos Vici Johnstone

This 1957 Pioneer JB superbly restored by Lowell Boyd was IEL's crowning achievement, a full-sized professional falling saw with direct drive. In it, the essential features of the modern chainsaw were brought together. *Collector Mike Acres, photo Lionel Trudel*

The Pioneer 650, a 1964 successor to the IEL model RA, was powerful and fast cutting. *Collector Mike Acres, photo Vici Johnstone*

The Pioneer 750, also released in 1964, was a larger variant on the 650, capable of handling a 32-in. (80-cm) bar.
Collector Mike Acres, photo Vici Johnstone

PIONEER

Pioneer

At the time of OMC's purchase of IEL in 1956, the Vancouver company was manufacturing three models: the HC, RA and JC. Except for a few parts that were die-cast, IEL was still making housings with sand-cast aluminum. OMC wanted to upgrade to die-casting and designed a new line of saws to be produced at its Peterborough factory for the 1958 season: models 400, 600 and 800. The 400 was an entirely new model designed for the occasional-user market, and the 600 was new but patterned after the very successful RA. The 800 was the JC with a few minor changes, a 7.5 cubic-inch (123-cc) west coast falling saw that sold to a narrow professional market, and was still sand-cast and built in quantity in Burnaby before the factory closed.

Improvements to the 400 and 600 came with the labels 400A and 600A, and later the 410 and 610. The next model year the 410 became the NU-17, and the 6-20 replaced the 610. New models for west-coast use were also on the drawing board, and in 1963 the 700 was introduced, followed by the 700G, a gear-drive version. The 1964 lineup offered models 450, 550, 650, 750 and 850. Pioneer scrambled to come up with the 11-10 in 1965, a close imitation of the new ultra-light Homelite XL-12. The 11-10 was the first of a long run of 3.0- and 3.56-cubic-inch (49-, 58-cc) XL-12 look-alikes. The 4.0-cubic-inch (66-cc)14-10 also appeared, the first in a successful series of professional models ending with the 4.7-cubic-inch (77-cc) 3270SC in 1973.

Lighter occasional-user saws were all the rage in 1971, so Pioneer contracted Oleo-Mac of Italy to build a version of its small saw and called it the P11. The Holiday II, a new occasional-user saw of 3.14 (51-cc) cubic inches was also introduced in 1971, starting a series of saws (including the 1072, 1073, 1074, 2071, 2073, P20, P25, P26 and P28) that continued until Pioneer ended production. The P10, which appeared in 1973, was manufactured for Pioneer by Quadra Manufacturing in Trail, BC, giving Pioneer a 2.1-cubic-inch, (34-cc) occasional-user saw of top quality and performance built in Canada. A new series of professional saws was introduced in 1973 starting with the 4.0-cubic-inch (66-cc) P40, to which the P50 was added in 1974—the first 5.0-cubic-inch (82-cc) saw Pioneer dealers had for quite some time. Then came the 6.0-cubic-inch P60 (98-cc) that was capable of handling a 36-inch (91-cm) bar, ideal for west coast logging and clearing. These models re-emerged with improvements as the P41, P51 and P61.

In 1979 Pioneer introduced the "Farmsaw," essentially the P41 with some modifica-

tions to reduce the cost and appeal to occasional users who wanted a saw with a 24-inch (61-cm) bar. It was a great success, selling in the tens of thousands. The paint job was orange and grey, a departure from the regular Pioneer colour scheme of yellow and green. The following year the "Farmlite" was introduced with a 3.5-cubic-inch (37-cc) engine, but it did not sell nearly as well as the Farmsaw. When OMC decided to quit building chainsaws and concentrate on outboard motors in June 1977, Pioneer had a complete model lineup of saws that were selling well in most markets.

Harking back to IEL, Pioneer employees formed a group to take over the operation, but were unable to return to production until May 1978. Distributors and market share were lost, placing the new company behind the eight ball from the start. Pioneer soldiered on with an excellent set of saws. It was even testing the new P35 model, designed to compete with the hot new European imports from Stihl and Husqvarna, but the company was underfunded and unable to recover its place in the market. It lasted until 1983, when it was forced to sell out to Electrolux.

The Pioneer 450 was the 1964 entry in a new series started by Pioneer for the casual-user market. *Collector Mike Bjellos, photo Vici Johnstone*

Right: The 1969 Pioneer 1200 was the final version in a series meant to compete with the Homelite XL-12, an ultra-light saw aimed at both the home and professional markets. *Collector Mike Acres, photo Vici Johnstone*

Bottom right: The 3270SC, a mid-size pro saw with a 77-cc engine. A good workhorse saw that sold very well. *Collector Mike Acres, photo Vici Johnstone*

Above: The Pioneer 2400 was a 77-cc model released in 1968. *Collector Mike Acres, photo Vici Johnstone*

Above left and above: The 1979 Pioneer Farmsaw, aimed at the non-professional user who wanted some power. Thousands were sold and are still cutting firewood decades later. *Collector Mike Acres, photos Vici Johnstone*

Left: Pioneer's valiant final effort, the 1982 P35 had all the mod cons—anti-vibe mounts, chain brake, quiet muffler, snappy design. Never marketed, it was being tested when the firm closed. *Collector Mike Acres, photo Lionel Trudel*

PIONEER
PARTNER

Pioneer/Partner

Electrolux was already operating a factory in Huron Park, Ontario, where it was building the former Frontier line under the name EMAB Canada (Electrolux Motors AB). Here Pioneer continued, but under a new brand name: Pioneer/Partner. The models that were initially manufactured at Huron Park were the P42, P52 and P62, with subsequent additions of the latest Frontier models: the 330, 350, 360 and P39, which was a redone Farmsaw II. Later, a P45 was created by increasing the displacement of the P42 to 72 cc from 65 cc and adding a finger port in the cylinder wall opposite the exhaust port. The P62 became the P65 after some slight improvements.

From the Partner models made in Sweden, the 400, 450, 500, 550, 5000+, 650 and 7000+ were added to the Canadian-made saws for a complete range of models and prices, all marketed under the brand name Pioneer/Partner. The product line continued until 1988 when the next name-change took place and the saws became Poulan Pro. The Canadian factory kept operating under this brand until 1993, when it ultimately became a casualty of the Mulroney government's Free Trade Agreement. Its operations moved to Shreveport, Louisiana, marking the end of the great Canadian chainsaw story that began in Vancouver during the early years of World War II.

The 34-cc 1984 Pioneer/Partner 330 had its roots with another Canadian company, Frontier, whose Ontario plant had been acquired by Pioneer's new owner, Electrolux.

HORNET

Hornet Industries

After he left IEL in the closing years of World War II, pioneering Vancouver sawmaker D.J. Smith moved to Guelph, Ontario and formed Hornet Industries Ltd. Hornet immediately set to work building a reasonably light saw with a 6.9-cubic-inch (113-cc) engine. The model HJ was on the market by 1946, complete with Hornet-made bar and chain. An idler-type outer end handle was still in use at this time and Hornet offered bars up to 60 inches (152 cm). The chain was of the standard (scratcher) type but with 5/8-inch pitch (15.8 mm).

Smith also started investigating designs for a one-man saw that would sell in the Scandinavian logging market and although no example is known to exist, Hornet built enough units to be offered for sale in Finland. By 1948 he was offering the one-man DJ-3500H in Canada and elsewhere. This saw provided the inspiration for several European manufacturers and resulted in the Be-Bo, the first saw manufactured in Sweden. The first Partner saw was also patterned after the Hornet.

An improved HJ was also introduced at this time as the Model D. Hornet designed its own chipper chain and called it Planer Chain. It was available to fit all Hornet saw models along with standard chain and a newer scratcher chain called Shark chain. The last Hornet model to be released to the market was the one-man DJ-3600H, which had an automatic rewind starter.

Mall Tool Company bought out Hornet Industries and continued to operate the Guelph factory, building the DJ-3600H saw. Eventually the factory was closed and some staff elected to take their skills to the Mall factory in Toronto.

The last Hornet model to be released to the market was the one-man DJ-3600H, which had an automatic rewind starter.
Collector Mike Acres, photos Vici Johnstone

Right: The Ontario-built 1948 Hornet DJ-3500H inspired Sweden's first saw, the Be-Bo. *Collector Marshall Trover, photos Brian Morris*

Below: The1946 Hornet HJ was a reasonably light and well-designed one- or two-man saw. *Collector Mike Acres, photos Lionel Trudel*

PM (Power Machinery Ltd.)

Power Machinery Ltd. was part of Vancouver's wartime manufacturing boom. Incorporated in 1943, its first major production was a table saw. But by 1947 it had hired designer Sandy Megaw, formerly with IEL, and issued its first chainsaw, the Universal—designed for one-man use but also supplied with a helper handle. They subsequently increased the size of the engine to 90 cc (5.5 cubic inches), added an automatic rewind starter and issued the result as the Universal A.

Weighing 36 pounds (16.3 kg) with a 20-inch (50-cm) bar, the early Universals were distinctive-looking saws with their unpainted aluminum finish. There were not many one-man saws on the market, and PM found that they were selling to customers all over the world. They painted their next run of saws red and gave them the name Woodboss. By 1950, PM had sold 5,899 of the first Universal, 2,499 of the A and 9,500 of the Woodboss—almost 18,000 saws in three years, not at all bad for a small Canadian manufacturer.

PM continued their new red colour scheme with the Redhead in 1950, which they soon redesigned with a somewhat larger engine capable of handling a bar up to 60 inches (152 cm) with a helper handle to adapt the saw for two-man operation. In these years when even one-man saws weighed well over 30 pounds, this was a successful strategy, so the Redhead was soon followed by the similar Torpedo. However, by 1952 the trend clearly was moving towards one-man saws, so in 1952 PM introduced the Rocket, fitted with a maximum 36-inch bar on a lightweight 24.5-pound (11-kg) powerhead. Chainsaws had still not made the breakthrough into

The New P.M. ROCKET-K1

A Man Can't Work This Way But this Saw CAN!

THE P.M. ROCKET CUTS AT ANY ANGLE

direct drive, so Sandy Megaw saved weight in this design by using a Gilmer belt reduction system instead of gears. PM kept producing Rockets until 1955, switching over in the later models to the Tillotson H diaphragm carburetors that enabled the saw to be run in any position.

It was a time of great ferment in the small-engine field, and in 1956 PM introduced its first direct-drive saw, Model 19. Even with its 6.25-cubic-inch (102.4-cc) engine and a 16-inch (40-cm) bar and chain, the unit only weighed 27 pounds (12 kg).

PM had cultivated a world market, but had its eye on central Canada's huge pulp industry. The IEL Beaver and other one-man saws had taken this lucrative market by storm and PM head Lawrence Killam wanted a piece of the action. He asked Sandy Megaw to design a line of saws for pulp cutting which would be branded "Canadien" in hopes of appealing to French Canadian pulp cutters, who outnumbered all others. Megaw used a 5.0-cubic-inch (82-cc) engine as the basis for a new direct-drive saw, the PM Model 21, which took up to a 30-inch (76-cm) bar with a 20-pound (9-kg) powerhead, and Model 22, a gear-drive saw which took up to a 48-inch (121-cm) bar with a 24-pound (10.8-kg) powerhead.

PM put all its resources behind these new saws. They hired outdoor personality Nels Adair to introduce the Canadien line across the country, signed up a network of new dealers and enjoyed high sales, especially in Quebec. They also sold well in Australia, New Zealand, Hong Kong, South Africa and the British Isles, taking advantage of preferential treatment accorded members of the British Commonwealth.

PM's designs stagnated for a time as they coasted on their success, and might have declined, but fate dealt them a favour. When crosstown rival IEL, by now Canada's largest chainsaw manufacturer, was sold to the Outboard Marine Corporation (OMC) in 1956, OMC moved the IEL factory to Peterborough, Ontario. This released a huge number of skilled chainsaw engineers, technicians and staff into the Vancouver job market and PM headman Lawrence Killam took full advantage. Early in 1959 Killam hired ex-IEL man Jack Stainsby as his chief design engineer to replace the departed Megaw.

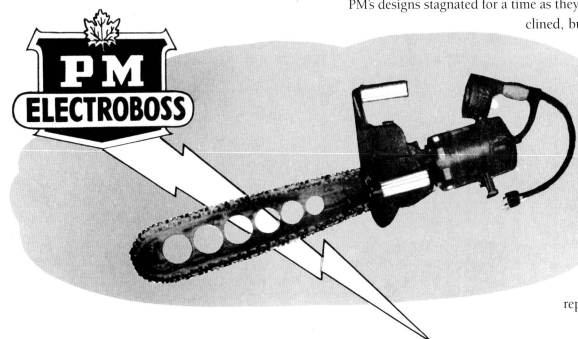

PM ELECTROBOSS

Stainsby went to work designing a new saw aimed at the professional market. It was a project that became more and more viable as more and more ex-IEL employees brought their expertise to PM. With this combined talent, PM was able to cut costs by manufacturing a range of parts, such as clutch drivers, that they had previously bought elsewhere. They gradually developed a highly motivated workforce, largely of former IEL employees who didn't want to forsake Vancouver for Ontario and, as the *British Columbia Lumberman* reported, "a number returning from Peterborough to lowlier jobs with the coast firm, foreman taking on drill-press work, for instance, in order to stay at the coast."

In March 1961, PM unveiled its new Model 270 Canadien at the Canadian Pulp and Paper Manufacturer's Convention in Montreal. Built largely of magnesium, the 270's powerhead weighed only 20 pounds (9 kg). It was a direct-drive saw that could take up to a 36-inch (92-cm) bar with its 95-cc (5.8-cubic-inch) engine. Although attractively designed with its cream-coloured body and metallic blue top, the 270 was meant to be, in Stainsby's words, "a chainsaw that the professional user wouldn't curse a month after buying." Indeed, although the 270 wasn't the cheapest, lightest or most powerful saw on the market, PM's new staff had also brought with them the rigorous IEL practice of field-testing their prototype saws before going into commercial production, and soon the Canadien's all-around proven quality won it fans around the world.

The 271, a gear-drive version, was added soon after to satisfy the needs of a shrinking group of professional users. This began a golden era for PM. The Finnish forest service tested and approved the saw, and suddenly Finland became a major market. Not surprisingly, US customers did not scramble to buy the 270 and its successors, nor did Germans—but Australia continued to be an enthusiastic customer, and so did Quebec. "If you get enough of these small customers throughout the world," said the new general manager, ex-IEL president Chuck Pulham, "you can get a nice business going."

However, the demands of international success were straining the capacity of PM's Commercial Drive factory on Vancouver's east side. Daunted by the expense of moving the operation, in 1962 Lawrence Killam accepted the offer of an English firm, Bristol Aero Industries of England, who bought the company and in 1964 moved the factory to a large hangar at the Vancouver airport. Bristol Aero kept the PM name and continued the Canadien line throughout the 1960s.

PM also continued to design new models, introducing the 275, a 7.5-cubic-inch (123-cc) direct drive for big timber work. Then it added the 276, a gear-drive version. Soon after that, the 175 was brought to the product line as a lower-priced unit using the 270 chassis and a slightly smaller engine. Two years later came the 177 and 187, newly designed units still using the 5.8 (95-cc) cubic-inch engines but weighing 3.5 pounds (1.6 kg) lighter than previous models. These units kept sales rolling along for PM and widened the

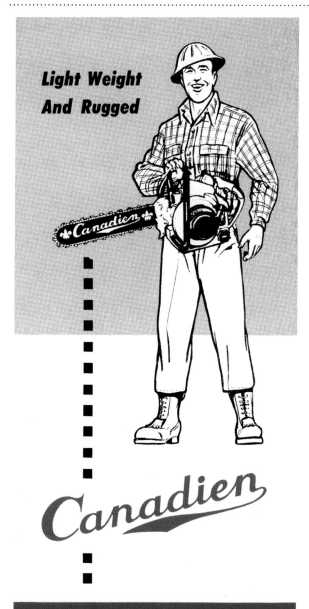

Light Weight And Rugged

Canadien

for BIGGER PRODUCTION ... BETTER ECONOMY

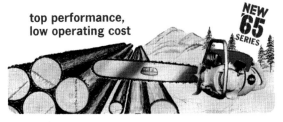

top performance,
low operating cost

NEW '65 SERIES

CANADIEN 270 DIRECT DRIVE Chain Saw

The "Canadien" brand became so well known, people forgot the company's actual name was PM, just as people forgot IEL was the parent of Pioneer.

product line. The final editions introduced by PM were the 330, a 3.4-cubic-inch (56-cc) lightweight unit, and the 340, its 4.2-cubic-inch (69-cc) big brother. These saws weighed in at 13.5 pounds (6 kg) for the powerhead.

PM's saws were so successful that the company became known as "Canadien," just as IEL was effectively "Pioneer" long before the company actually changed its name. But in 1969, Bristol Aero sold the chainsaw operations to the Skil Corporation of Chicago. PM's engineering department had been hard at work on an even lighter occasional-user model, which may have been one of the factors that attracted Skil to purchase PM, and then release the small saw as the 944. Within a year, Skil decided to move saw manufacturing to its main factory in Chicago. Many skilled workers at the Vancouver plant had the option of moving, but few if any chose to accept it. In fact, even though the plant's general manager and Skil vice-president Jim Hutchinson was an American by birth, he resigned from the company rather than make the move. Instead, Hutchinson got together with eleven other ex-employees and schemed to form a new BC-based chainsaw company, Quadra Engineering, creators of the Frontier brand.

The 1949 Woodboss, second version of the one-man saw that launched Vancouver-based Power Machinery (PM). Low price helped it sell in decent numbers despite mechanical shortcomings. *Collector Mike Acres, photo Lionel Trudel*

Above and above right: The 1950 Redhead, equipped with a "one-man/two-man" rear handle system and powerful engine, could be used in all sizes of timber. *Collector Mike Acres, photos Vici Johnstone*

Below and right: The 1952 PM Torpedo was a lightweight two-man saw produced at a time when the two-man saw was out of date. *Collector Marshall Trover, photos Brian Morris*

Left and top: PM's 1956 19A Pacemaker, with a good-sized 102-cc engine and all-position Tillotson carb, proved to be a powerful, versatile saw. *Collector Mike Acres, photos Vici Johnstone*

Above: Battle-scarred 1961 Canadien 210. *Collector Mike Acres, photo Vici Johnstone*

The 1961 Canadien 270 earned a reputation for power and reliability in a highly competitive environment, launching PM into the pro saw market in a big way. *Collector Art Patterson, photo Lionel Trudel*

FRONTIER

Frontier (Quadra Manufacturing Ltd.)

Quadra Manufacturing Ltd., the last significant power-saw manufacturer to be based in British Columbia, was the result of Jim Hutchinson and the other eleven ex-employees of Skil scrimping, saving and working at other jobs while they looked for a way to get their newly formed company off the ground. Meanwhile, they discussed and drew up plans for a saw that would stand out from the competition. They also learned there were considerable local and federal start-up funds available—if they would consider moving their venture to Trail, a smelter town in the BC interior 390 miles (628 km) east of Vancouver. Apart from government funding, there was enthusiasm on a grassroots level—240 Trail citizens did their own fundraising and formed Kootenay Incentives Ltd. to invest in Quadra. By September 1972 Quadra had opened a Trail factory and was producing its first Frontier chainsaw, the Mark I.

Aimed squarely at the consumer market, the top-handle Mark I had a petite 6.6-pound (3-kg) , 2.2-cubic-inch (36-cc) powerhead and was designed to use 1/4-inch (19-mm) pitch chain. However, it could take up to a 16-inch (40-cm) bar, and Quadra found a brisk trade with professional arborists and pulpwood cutters as well as with occasional users. Five months after the first Mark I, the company had sold ten thousand chainsaws and was already claiming a profit. In November 1974, Quadra had a total staff of "Two Hundred and Fifteen Artificers of Fine Chainsaws."

Frontier was going great guns during the 1970s, but began running into money problems. Their die-casting source was in Ontario, and most other suppliers were either in eastern Canada or the USA, which increased manufacturing costs. Their markets were in the east and in other parts of the world, increasing sales costs. The city of Trail also happens to be home to the world's largest lead, silver and zinc smelter and as a result has a strong union orientation; Quadra had to pay higher wages than other manufacturers, another contributing factor to higher costs.

Although Frontier offered several variations on their F35 model in 1979, and added the F48—a larger-engine-sized model made for them by Jo-Bu in Norway—the squeeze was too great and the company was placed in trusteeship. Operations were subsequently moved to an abandoned aircraft hangar in Huron Park, Ontario and carried on for a while as Trail Manufacturing Ltd. (TML). Frontier had been manufacturing saws for Electrolux, who in turn sold them under its brand names Husqvarna, Jonsered and Partner. When the TML operation got into trouble, Electrolux bought the company to protect its source of small saws, and continued manufacturing in Huron Park under the Electrolux division EMAB Canada, which finally ceased operations in 1993.

Top and above: From a cold start in 1972, by 1974 Trail, BC-based Quadra Manufacturing had sold 10,000 units of their first model, a petite 6.6 lb. (3kg) top handle design that found buyers among professional arborists and pulpwood cutters as well as occasional users. *Collector Mike Acres, photos Vici Johnstone*

FRONTIER
The Small Saw with Guts

The 1946 Precision Model 1 was an early one-man saw used primarily in the east and south to cut smaller timber and pulp wood. *Collector Marshall Trover, photos Brian Morris*

PRECISION

Precision Parts Limited

The Precision chainsaw appeared in 1946, manufactured by a new company headed up by Alastair Grant in Montreal, Quebec. Precision made one basic engine unit, the "Precision One Man," and offered different types of cutting attachments and handles to suit. To access the US market, Precision initially had a sales office in Rouses Point, New York and later moved to Danbury, Connecticut, where it expanded the office into a US assembly plant.

Precision's first saw was the Type 1, which featured bow-type cutting attachments in sizes of 14, 18 and 24 inches (36, 46 and 61 cm), and the cage-type handle system with a "belly pan" so the operator could use one's stomach or upper legs to help position the saw. With some modifications, the Type 1 remained in production until about 1952: its Series 1 variation had a separate clutch lever, Series 2 had a combined twist-type clutch/throttle control, and Series 3 was the result of an attempt to remove weight from the fuel tank/blower housing.

The Type 2 Series 1, introduced in 1946, had the same engine as Type 1 but featured a 22-inch (56-cm) beaver-tail guide bar, and one-man handles instead of the cage handle arrangement. In 1952, 19-inch (48-cm) and 27-inch (69-cm) bar lengths were introduced on the Series 2, which was sold through 1955. Type 3 Series 1—sold from 1946 to 1950— offered a two-man arrangement of either 20-, 26- or 32-inch (51-, 66- or 81-cm) guide bars with a helper handle on the outer end and a manual clutch with separate engagement lever.

In 1952 the Model A appeared with automatic rewind starter, offered in either the bow-saw or beaver-tail-bar version, but stopped being built in 1953. Also in 1953, Precision's final models were introduced with a lighter-weight blower housing and cylinder cover, and a stamped steel fuel tank mounted on the rear of the handle cage. The Model 18 was the bow-saw version, with holes drilled in the bow housing to remove some weight, and the Model 19 was the beaver-tail version. Production of both ended in 1955.

Precision carried on in the chainsaw business, distributing brands such as Hoffco, and the Comet Diesel from Norway, to sell in Canada and the US.

Precision Model 19, an updated 1951 version of the Model 1, smaller, but with no major mechanical changes.

Collector Marshall Trover, photo Brian Morris

Homelite gave rise to a whole new era in chainsaws when it unveiled its 12 lb. XL-12 in 1964 and followed it with a whole galaxy of models that set the pace for lightness and toughness.
Collectors Art Patterson (blue) and Lowell Boyd (red), photo Lionel Trudel

5

The Industry Expands

I n the USA, Titan, Mall and Disston were familiar names by the war's end, but the industry was already spawning dark horses who were on their way to overtake them. For several decades after World War II, literally scores of chainsaw makers blossomed all over the world. By 1957, there were at least forty-five manufacturers worldwide, cataloguing over two hundred models of saws.

In North America alone, several dozen manufacturers sprang up in the 1950s, offering casual-user or "farmer" saws built around either a Power Products or a West Bend engine. The list of the decade's manufacturers, mostly from the USA, includes Strunk, Skarie, Lebanon, Indian, Tree Farmer, David Bradley, Lancaster, Hoffco and England Motors (based not in England but in Pine Bluffs, Arkansas). The list also includes Monark Silver King of Chicago (not to be confused with the Swedish firm Monark-Crescent), who in 1957 offered their Model 47 that, although too heavy to be used with one hand, displayed an overhead-handle design that predated the one-handed "arborist" saws of later years. As with most aspects of the post-WWII world economy, business in the United States was booming.

The Poulan line had an enthusiastic booster in the person of logger sports champion Ron Hartill.

Texan Claude Poulan reinvented the bow saw for cutting the small timber in the southern US forests. *Collector Marshall Trover, photos Brian Morris*

POULAN

Poulan

Claude Poulan (pronounced "Polun") began learning the logger's trade at an early age in the pine forests of northern Louisiana. In 1944 he was working for International Paper in East Texas, supervising pulpwood logging in the area's vast forests of hardwood and loblolly pine. Poulan's crews were made up of German POWs using two-man chainsaws that were state-of-the-art equipment for the time.

Most of the sawing consisted of bucking the logs into short lengths and for this, the two-man saw was very ill-suited. It was oversized for the work and with the helper handle in the way it was hard to get it down low enough to finish the cut. The wide bars of the time with their bulky scratch chains were also very susceptible to binding as the log shifted, and Poulan observed that it was usually the top of the bar and the non-cutting returning loop of the chain that became bound. Poulan's crews employed a third worker to position the logs with a pry bar and he considered the system woefully inefficient.

Poulan set out to develop a saw better adapted to small-log cutting. Legend has it that he hammered together the first working model using a castoff Chevrolet truck fender. Instead of running the chain around a solid bar, Poulan re-routed it around an open loop broader in diameter than most of the area's trees. In use, only the lower loop with the cutting part of the chain touched the wood; the back loop stayed clear. Since the cutting part of the bar could be very narrow, it was less susceptible to getting pinched in the cut. This cutting portion of the loop was also designed with a pronounced "bow" so that it could be pushed completely through the cut without hanging up like the prevailing straight bar with helper-handle did. Because the upper loop was shielded, it was also safer to use.

The idea had been tried before by Arsneau, and Stihl had produced a bow saw, but Poulan introduced it in the right place at the right time. When the war ended, he went into business for himself, working out of his garage in Alto, Texas, an area distinguished by the Davy Crockett, Sabine and Sam Houston National Forests. By then, he had already built a successful business adapting the bow guide and selling it to Mall, Disston and other chainsaw makers.

But Poulan was already designing his own chainsaw. Moving his operations first to Marshall, on the Texas side of the highway between Dallas and Shreveport, Louisiana, and then to Shreveport itself, Poulan filled orders for his bow guide under the name Ark-La-Tex Bar Service while all the time designing his first prototype. Using an engine designed primarily for generators, Poulan fitted a chain and his bow attachment and in 1946, premiered the first Poulan chainsaw, the two-man Model 2400. Establishing his own foundry for casting parts took up too much room in the 4000-square-foot Shreveport building, but the success of the 2400 enabled Poulan (joined by among others Ernest Garrett, who designed an assembly line for Poulan saws, and Claude's brothers Harry and Fletcher) to move to a new twelve-acre site in Shreveport, which eventually housed a 55,000-square foot factory. All four of these chief executives were hands-on proponents of their products, always ready to lug saws out into the bush to demonstrate them to loggers on the job.

After Fletcher Poulan joined in 1951, in short order the company produced eleven different models of saws and expanded their sales network across the country. They went public, selling shares on the stock market as the Poulan Manufacturing Corporation. Such was their success that in 1959, after two years of negotiations, all of the stock of Poulan Manufacturing and its subsidiary, the Poulan Saw Company, was bought up by Charles T. Beaird.

Beaird was a Marine captain during World War II and a scion of his family business, the metal fabricators J.B. Beaird Company. He closed the deal with Poulan in 1959 and with Beaird's interest and investment, the new Beaird-Poulan Inc. doubled in size and the company joined the front rank of the world's chainsaw manufacturers. It enjoyed two more successful decades, being sold to Emerson Electric in 1973 and merged with Weed-Eater. The new company operated as Poulan/Weed Eater and grew considerably in size before being bought by Electrolux in 1978.

Above and below left: Thanks to the success of the early Poulans, bow saws became standard in small-timer country.
Collector Marshall Trover, photos Brian Morris

Above: The early Poulan models used Homelite power. The narrower cutting bar of the bow assembly reduced the problem of pinching or binding that plagued the small timber bucking process. *Collector Marshall Trover, photo Brian Morris*

Left: This 1953 Model A was the first Poulan one-man with a complete package designed from scratch. *Collector Marshall Trover, photo Brian Morris*

Left: The 1976 Super XXV (Micro 25) remained in production until the mid-1990s. *Collector Mike Acres , photo Vici Johnstone*

Inset: In 1964 Poulan was still making successful gear drives that worked equally well with bar or bow. Rugged and a little heavy, this Model 40 was not popular with the pro fallers up north. *Collector Marshall Trover, photo Brian Morris*

Below: The Model F200, a very rugged model designed for rough use, was another success in Poulan's line of bow and bar saws. *Collector Marshall Trover, photo Brian Morris*

LOMBARD

Lombard

For years Lombard claimed to be "the oldest name in chain saws" and for what it's worth (since they by no means invented or manufactured the first chainsaws) that was apparently true until Husqvarna entered the market in 1959. Lombard was founded in 1894, and Lombard researcher Tom Hawkins has found records of the Lombard Governor Corporation—making governors for hydraulic water wheels—dating back as far as 1902.

Lombard joined the World War II chainsaw boom in 1943, when company owner Henry Warren made the decision to begin building saws at his factory in Ashland, Massachusetts. He got into the business with a vengeance, with the ES electric, Model PS pneumatic and the Model GS gas-powered saws. For the GS, Warren used a 6-hp Homelite 24X1 engine that had originally been designed to drive water pumps and electric generators. All of Lombard's saws from 1943 to 1948 came with the same standard 24-inch (61-cm) bar and chain combination, with 36-inch (91-cm) or 48-inch (122-cm) bars available for the gas and electric units.

Lombard sold five hundred saws in its first three years; not bad considering that these

The 1949 OMS, the first one-man saw manufactured by Lombard, used an updated Homelite model 20X1 pump/generator motor. *Collector Marshall Trover, photo Brian Morris*

were still heavy two-man saws aimed at the professional market. In 1946 they moved up to the Homelite 24X2 engine on their gas saws at the same time that Poulan adopted the 24X2 for use on their 2400 bow saw. In 1948 the 24X2 was re-rated from 6 to 7 hp, so all of Lombard's previous gas saws became known as GS(6), and their new 1948 model, with automatic oiler and centrifugal clutch, was designated the Model 7, and could be fitted with a bar as long as 60 inches (152 cm) or even 72 inches (182 cm).

Lombard introduced its first one-man saw, the (Model OMS) in 1949. Using a Homelite 20X1 engine, the OMS weighed only 35 pounds (16 kg) with a 19-inch (48-cm) bar and chain. But it was the end of Lombard's association with Homelite; they established themselves with a vengeance in the emerging lightweight saw market with their next one-man saw, the Woodlot Wizard in 1951. The Wizard weighed only 28 pounds (13 kg) and used a Power Products AH-47 engine.

However, the market was still basically loggers and even as the one-man saw changed the way things were done in the woods, thirty-odd pounds is still a lot of weight to heft around all day. Even though the longest bar offered on the OMS4 was 34 inches (86 cm), it featured a slotted tip for a snap-on helper handle. The OMS 42 also had a two-man configuration.

Along with this diversification of models, Tom Hawkins points out that Lombard saws were offered with a variety of chain: the traditional scratch chain, the up-and-coming Cox chipper chain, and Warren chain. The Power Products relationship worked out well with Lombard. In the late 1950s, using the PP AH-81 (7.98 cubic inches/131 cc) engine, they created the Fury saws, a favourite among logger sports contestants that Hawkins calls "the best competition saws in the world at the time; their dominance lasted into the early 1960s." By this time, after decades of offering not only governors but also chainsaws, plastic moulding machines and contract machining services, the name Lombard Governor Corporation didn't really do the firm justice, so in autumn 1962 they incorporated as Lombard Industries Inc. A year and a half later, they sold their chainsaw division to the American Lincoln Corporation (ALC), and the manufacturing operations moved from Ashland to Toledo, Ohio.

Someone at Lincoln must have been a chainsaw enthusiast, because the chainsaw division thrived under ALC. In 1965, featuring its own Lombard 4.2-cubic-inch/69-cc engine, the company introduced the AL-42. Although the name Lombard was prominent on the bright red body, the cream-coloured sprocket cover sported the American Lincoln name and the proud boast "The World's Most Powerful Lightweight Chain Saws." Modelled on the popular Homelite XL12, the AL-42 began a series of successful saws aimed at the booming consumer market.

Amid the grow-or-die trend of the corporate world, by 1968 ALC was one of twenty-

The 1946 Lombard GS(6), a heavy two-man saw powered by a Homelite 24X2 pump/generator motor. *Collector Marshall Trover, photo Brian Morris*

Above and below: For 1960, the 650 Wonder looked old-fashioned with its external fuel tank, but in other respects—direct drive, all-position handling, feathery 16-lb. (7-kg) weight—it was right up to date. *Collector Mike Acres, photos Vici Johnstone*

two divisions of Scott & Fetzer Company of Lakewood, Ohio, and in 1970 itself split into three divisions. The new Lombard Power Equipment division was based in Cleveland, but the chainsaws were manufactured in Montreal until early 1973, when Scott & Fetzer's Campbell Hausfeld division moved chainsaw production back to Ohio and it opened a new plant in Harrison.

The early seventies were the beginning of the end for many of the world's smaller chainsaw manufacturers, and Tom Hawkins is critical of Campbell Hausfeld's management of the Lombard line. "From 1973 to the final Lombard (about 1983)," he writes, "designs remained the same (just a change of paint and model number), some going as far back as 1966." Hawkins also points out that he found "little or no advertising of Lombard from 1973 to 1979."

In 1979 a group of Harrison private investors banded together as the American Power Equipment Company, bought Lombard and tried to reinstate it as a force in the US chainsaw industry. But its time had passed. Having lost its edge in a competitive market, Lombard made its final push in a time when chainsaw sales were in a slump. Its ascent was further blocked by the rising dominance of huge manufacturers such as Stihl and Electrolux/Husqvarna. By 1984, Lombard had ceased production.

Above: The GS(6) had a manual clutch and an oiler built into the tailstock. Even for the time, it was a bit of a dinosaur. *Collector Marshall Trover, photo Brian Morris*

Below: The 650 Wonder was a good example of the low cost chainsaws sold to farmers and occasional users in the 1960s by dozens of different manufacturers. *Collector Mike Acres, photo Vici Johnstone*

more cutting for your dollar

NEW
HOMELITE
EZ

DIRECT DRIVE CHAIN SAW

This new Homelite EZ chain saw combines all these features to give you more cutting for your dollar - full 5 horsepower, new light weight of only 19 pounds and new low cost. In no other direct drive chain saw will you find so many features you want for any type of cutting. Look inside for the many ways you can get faster, easier, more profitable cutting with the new Homelite EZ.

CONVENIENT TIME PAYMENT PLAN

Feel the FLOATING POWER in a Free Demonstration

Homelite didn't come on the chainsaw scene until 1949, but with their long experience in two-cycle engine-making, it wasn't long before they were producing saws of professional quality.

HOMELITE

Homelite

The name Homelite is rightly associated in many people's minds with lightweight chainsaws for home use—but when Charles H. Ferguson founded the Home Electric Lighting Company in Port Chester, Connecticut in 1921, his aim was to make lightweight generators for home use. Homelite didn't make its first saw until 1949: the one-man, 30-pound (13.6-kg) 20MCS. Like most first attempts it had some bugs, but these were addressed in the 1951 26LCS, which met with considerable success. The next saw, the 1953 5-30, used die-cast magnesium parts and the Tillotson H-series all-position carburetor, making it a very competitive machine for the time. The 1954 Homelite 17 weighed only 20 pounds (9 kg), less than the McCulloch 47. They also made larger logging saws with 129-cc (7.88-cubic-inch) engines and up to 72-inch (182-cm) bars, the 7-29 in 1956 and the 8-29 in 1957.

In 1955 Ferguson sold out to the budding conglomerate Textron Inc., which continued to expand the business. The direct-drive EZ appeared in 1956. It had a 5.01-cubic-inch engine and used the same basic design as the earlier 5-20. Many models followed on into the mid-1960s using the same design, finishing up with the 775D and 775G. There were models called ZIP, WIZ and

HOMELITE
SUPER
XL

166

BUZ, Homelite's first true consumer chainsaw. Between 1959 and 1965 the company also produced large professional saws beginning with the 9-23. Up to this point Homelites had upright designs with a rounded profile and red and green colour schemes.

The new "C" series beginning in 1962 had horizontal cylinders and a low, rectangular profile. The metallic-blue, direct-drive C-5 had a 4.71 cubic-inch (77-cc) engine, followed in 1963 by the C-7 with a 4.91 cubic-inch (80.5-cc) and then the C9 with 5.22-cubic-inch (85.5-cc). In 1965 the XP-1000 was added to the line, offering a 6.1-cubic-inch (100-cc) engine and bar lengths up to 30 inches (76 cm). A gear-drive version, the XP1100, could handle bars up to 60 inches (152 cm) for the big timber fallers. The Super 2000 followed with a 7.0-cubic-inch (114-cc) cylinder. There were various other versions of direct-drive and gear-drive saws that continued in production through the mid-1970s using the same chassis.

In 1964 Homelite set the chainsaw industry on its ear with the XL-12, an ultralight consumer chainsaw with professional quality. The 3.3-cubic-inch (54-cc) powerhead weighed only 12 pounds 12 ounces (5.8 kg), establishing the company as a major manufacturer for the homeowner market. By 1971, when Homelite celebrated its fiftieth anniversary, it had sold three million saws; by 1978 it was shipping a million a year.

There were so many models branded "XL" with engines ranging in size from 1.6 cubic inches (26.2 cc) to 5.01 cubic inches (82 cc) that it is difficult to list them all. They were all very good saws with various features such as the automatic oiler and compression release, and eventually electronic ignition, anti-vibration systems and chain brakes, but by the early 1990s there were no more truly professional saws. In 1994 Textron sold Homelite to John Deere, which continued to focus on consumer models under 50 cc.

Above: The first Homelite saw, the 1949 20MCS, had magnesium castings, belt reduction drive, automatic rewind starter and a rear handle and float carburetor that swivelled for falling. It had a few bugs, but it was an impressive start for a company that would soon emerge as a major player. *Collector Marshall Trover, photo Brian Morris*

Left: The 1951 26LCS was an improvement over the 20MCS with slanted cylinder and cast alloy gas tank. *Collector Marshall Trover, photo Brian Morris*

Left: The 1956 EZ. *Collector Mike Acres, photo Lionel Trudel*

Below: The 4-20 for 1955 was one of Homelite's most successful small gear-drive models. Some were still in use 30 years later.

Collector Marshall Trover, photo Brian Morris

The 1956 EZ was the first direct-drive saw produced by Homelite. *Collector Mike Acres, photo Lionel Trudel*

The successful and durable C series was introduced in 1962. The C5G (top) had an outboard planetary transmission. The 1960 700GS (right) had a unique gear drive (inset) that could be shifted between two ratios. It didn't sell very well.
Collector Marshall Trover, photos Brian Morris

Above: The XL-1 was small, light, powerful and stood up to heavy use. *Collector Mike Acres, photo Vici Johnstone*

The Super XL 130 (left) and the XL 76 (below, left) were both mid-range workhorses with 58 cc displacement. The Super XL922 (below) was a 77-cc powerhouse that kept Homelite users coming back for more in this series. *Collector Art Patterson, photos Vici Johnstone*

If early Remingtons looked like Malls, it's because they were Malls. DuPont-owned Remington plunged headlong into the chainsaw business by purchasing the Mall Tool Co. in 1956.

Remington DUPONT

Remington: The Streamlined Chainsaw

The Remington Arms Company is the oldest continuously operating US manufacturer, and the oldest one that still makes its original product: firearms.

Although it was founded in Ilion, New York in 1816 as E. Remington and Sons, a maker of flintlock rifles, Remington's manufacturing output has been nothing if not diverse. Over the years it has made everything from ammunition to typewriters, and for a brief period it was even a chainsaw manufacturer—and an important one, at that.

During the Depression, Remington was purchased by the massive DuPont chemical concern, which during its long history has dealt in everything from gunpowder to nylon to plutonium. After World War II Remington's business flourished, but in the early 1950s the firearms market stagnated. However, Remington had enjoyed some success in developing powder-actuated tools, and felt that the tool business might be the place to expand. In 1956 Remington began its Power Tool division by paying $9.8 million to purchase the Mall Tool Company based in Park Forest, Illinois.

Announcing the **NEW** **REMINGTON** *Bantam* **CHAIN SAW**

Remington DUPONT

The Mall works became a separate enterprise under the Power Tool division. At the time Mall was offering a wide range of saws—gas, electric and pneumatic—and had been investing in an expensive switch from sand-casting to die-casting. Remington used its marketing expertise to create the Silver Logmaster and Golden Logmaster saws. In 1958 these models were replaced by the SL-5 and the GL-7, which were completely magnesium die-cast.

The success of these saws seemed to indicate that Remington was on to something good, so they made a big investment by hiring the famous Raymond Loewy to give a new shape to their saws. Loewy, called "The Father of Industrial Design," had made his name giving the "streamlined" look to everything from the Shell Oil logo to Greyhound buses. Using the same 5.0-cubic-inch (82-cc) engine that had been used in the Mall 1MG, GP, SL and SL-5, in 1961 Loewy came up with a radical new look for Remington's new Bantam (17 pounds/7.7 kg) powerhead. Loewy came as close as anyone has to turning a chainsaw into a modernist art object, but his design didn't handle well and had to be modified to improve its balance.

By 1962 Remington was offering a whole series of saws in the Bantam configuration. Some of them, such as the Bantam G and the Super 75G, were gear-driven saws for which there was still a market: the SL5-G had the transmission built into the crankcase in the conventional way, but the SL-5R and the GL-7R used outboard planetary transmissions.

Although it had made a lot of headway in the casual-user market, Remington's sales staff had been pressuring management to produce a saw specifically for the demanding conditions of the west coast logging industry. In 1963 Remington made a cautious entry into this market, combining new and old technologies to produce prototype saws in limited quantities. The direct-drive Super 880 and the gear-driven Super 880G had 5.8-cubic-inch (95-cc) engines with a new air-injection feature—a cooling fan that also pressurized the carburetor/air filter chamber to produce a supercharging effect. Their parts were sand-cast in the old-fashioned way, but their fuel tanks and air filter covers were made of Delrin, a new DuPont plastic. The saws held up well in field-testing, but the limitations of sand-casting made them a bit heavy—the 880G powerhead weighed 28 pounds (12.7 kg). A decade before, that would have seemed feather-light, but almost overnight, as one-man saws became the standard, even professionals were expecting that new saws would become lighter and lighter. Although a later saw, 1965's PL-6, used their cylinder design, the Super 880 saws themselves were not continued.

In the same year, 1963, in the west coast woods Remington tested a mystery saw about which little information remains. Like the 880s, it was a sand-cast prototype, a gear-driven saw with a very large and powerful 123-cc (7.5-cubic-inch) engine, and like the 880s, despite its effective performance it was considered too heavy to be introduced to the market.

Remington made some improvements to the Mall model GP to produce the 1956 Silver Logmaster. It was a stop-gap until the fully die-cast model SL-5 was ready. *Collector Marshall Trover, photos Brian Morris*

Top: Remington's 1958 SL-5 was a die cast version of the Silver Logmaster.

Bottom: SL-5R was a gear drive version with 3.56:1 reduction outboard planetary transmission. The oiler was unique in that it pumped the oil on the return stroke of the button.

Collector Mike Acres, photo Vici Johnstone

Remington decided to stay with the proven success of the Bantam shape and design, and with the prevailing strategy of making chainsaws as small and light as possible. Aware of the success of Homelite's diminutive XL-12, late in 1964 Remington introduced its PowerLite series with the 59-cc (3.6-cubic-inch), 12.5-pound (5.6-kg) PL-4. This was followed by the 4.0- cubic-inch PL-5 and the 5.7-cubic-inch PL-6, a fully professional model with a very similar cylinder design to the 880. The smallest of the PL family arrived in 1968. The SL-9 had a 9-pound (4-kg) powerhead and a 2.8-cubic-inch (46-cc) engine. A big brother was added a year later when the engine displacement was increased to 3.1 cubic inches (51 cc) and the name SL-10 applied.

Remington had entered a crowded marketplace, competing on one hand with established companies that offered quality and innovation, and on the other hand with countless smaller outfits who were quick to slap a bar onto a two-stroke engine and call themselves chainsaw makers. Nonetheless, it had made some significant machines, and established itself as a creditable manufacturer in the eyes of both professionals and casual users. During the 1960s, the word in the Power Tool Division was that with DuPont's virtually limitless capital for distribution, promotion and R&D, Remington had every intention of becoming the world's number one chainsaw maker.

However, Remington was a big company and at the same time that the Power Tool Division was celebrating its success and looking to the future, management were noticing that the long slump in firearm and ammunition sales was over—that in fact, the arms side of the business was booming. Someone at the top decided that the company should concentrate on the side of Remington's business they knew best, and in 1969 the Power Tool Division was sold.

The buyers were a venture capital group named DESA Industries Inc. DESA continued making various power tools under the Remington name, including the Power Lite gasoline saws, two electric models and two pneumatic models that had been made since the Mall years. Some new designs emerged from DESA/Remington, such as the 1971 Mighty Mite and the Mighty Mite Auto. Under various model names, DESA continued to issue a number of saws that had been introduced between the Power Lite series of 1964 and the changeover in 1969, but after only ten years they sold the gas chainsaw line to Alpina of Italy, and now only electric chainsaws appear under the Remington name.

Left and above: In 1959 Remington put the Mall look behind it with the futuristic new Bantam designed by Raymond Loewy, creator of the Shell Oil logo. *Collector Mike Acres, photos Vici Johnstone*

Below: The 1964 Remington 773. *Collector Mike Acres, photo Vici Johnstone*

Right and far right: The 1963 Super 880, Remington's first attempt to build a professional timber saw, had a new 82-cc engine with a "turbo" carburetor pressurization feature years ahead of its time. *Collector Mike Acres, photos Vici Johnstone*

Right and below: The 754 Army edition was produced by Remington in 1967 for use in Vietnam, where it was often seen clearing helicopter landing pads. *Collector Marshall Trover, photos Brian Morris*

Top left: Cutaway version of the 1965 PL-4, Remington's 12.5 lb. (5.7 kg) answer to the Homelite XL-12. *Collector Mike Acres, photo Vici Johnstone*

Top right and above right: The tiny 1968 Powerlite SL9, the world's first 9-lb. (4-kg) chain saw, very popular with arborists. *Collector Mike Acres, photos Vici Johnstone*

Left: Powered by the 82-cc engine first used in the Super 880, the 1966 PL-6 thrust Remington into the pro saw market. This saw could cut. (It barked at you.) *Collector Mike Acres, photo Vici Johnstone*

177

A 1972 Druzhba, Russia's national chainsaw.
Collector Mike Acres, photo Lionel Trudel

Europe Bounces Back

Adjusting to a New World

It took European chainsaw manufacturers a few decades to react to the post-war growth of the industry in North America, but in time they came back with surprising strength. At Stihl and Dolmar in the 1950s, the designers and engineers worked in an environment that was nothing like the relative isolation that Andreas Stihl and Emil Lerp had enjoyed thirty years before. The field thronged with competitors—in North America, in neighbouring countries such as Norway and Sweden, and in Germany itself—and if all of those sawmakers were in a sense the enemy, competing for the same customers and poised to capture the emerging homeowner/casual user market, they also offered a wealth of innovation and ideas that encouraged each of them to constantly modify and improve their designs.

Solo

While technical advances were being made in other parts of the world, Germany, despite the setbacks of military defeat, continued to produce important innovations. In fact, perhaps it was the harsh climate of food rationing, Allied occupation and general economic disrepair that stimulated certain individuals to thrive.

Take, for example, the brothers Hans and Heinz Emmerich. Before the war, they had worked on developing small two-cycle engines for model airplanes. After the war, while living in the town of Obertürkheim on the eastern bank of the Neckar River, they invented a backpack sprayer enabling basic crop-dusting to be reduced from a team of hand-sprayers to a single person. Soon "solo" motorized spraying became the industry standard and the Emmerichs chose as their new company name Solo Kleinmotoren GmbH. But cropdusters and many of their other farming products were hopelessly seasonal; with more than one hundred workers on the payroll, they needed sales all year round.

The Emmerichs saw a solution in the one-man chainsaw, more powerful and more versatile than ever thanks to such developments as chipper chain and all-position diaphragm carburetors. As yet, the concept of the small "consumer" saw had not particularly hit the European market, but it was a power tool ideally suited to the Solo specialty of small, lightweight two-cycle engines. In 1958 they released the Solo Rex as "Europe's first direct-drive chainsaw." Weighing only 26 pounds (11.4 kg) with a 17-inch (43-cm) bar and chain, the Rex used plastic parts more extensively than any saw before it.

Solo claimed their 1958 Rex was the first direct-drive chainsaw in Europe. They used injection-moulded and blow-moulded plastic parts years before other manufacturers. *Collector Mike Acres, photo Vici Johnstone*

Within a few years Solo had set up subsidiaries throughout Europe as well as in Ghana and the USA and by the 1970s, Australia and New Zealand, making everything from mist blowers to mopeds. By 1969 Solo were putting out fifty thousand chainsaws a year. In 1977 they signed a three-year contract to supply Husqvarna in Sweden with "consumer chainsaws" for casual users. Earlier in the 1970s, they had produced the lightweight Solo 600 with the new "arborist's" top handle design—virtually the identical saw to its Stihl contemporary, the 015 with an 8.1-pound (3.7-kg) powerhead.

In 1966 Solo introduced the "Combi" motor, essentially a lawnmower head that could be easily removed and used on other power equipment—there was even an outboard motor attachment. They introduced a multi-head chainsaw called the "Multimot" in 1983, the Solo 644 chainsaw in 1987 and the electric Model 614 in 1989. Solo have produced a range of saws in a range of sizes over the years, but they continue to be most strongly identified with the use of their "kleinmotoren" to produce lightweight, high-quality chainsaws.

Above: The 1974 616 had a sleek 45-cc design that brought lots of customers to Solo. *Collector Mike Acres, photo Vici Johnstone*

Below: The 1965 Solo Twin's unique side-by-side cylinders gave it lots of power and moderate success despite its extra weight. *Collector Mike Acres, photo Lionel Trudel*

SACHS
DOLMAR

Dolmar Goes East

Dolmar also experienced a hiatus in international sales and innovation during World War II, but introduced the one-man CP in 1952—a gear-driven unit that weighed about 30 pounds. With the diaphragm carburetor still in development, the CP had, like many other saws, a swivelling guide bar that changed angle for falling and bucking. But the whole look of chainsaws changed forever in the 1950s. In 1957 Dolmar unveiled the CF, advertising it with images of swooping jet planes to show how its new diaphragm carburetor would enable it to run in any position—even upside-down. It was still gear-driven, but even with the weight of the transmission, it weighed only 22 pounds (10 kg).

Dolmar thrived in the 1960s, but entered a fateful chapter in 1975. In that year, Fichtel & Sachs, a German automotive manufacturer, bought a controlling interest in Dolmar. Its main reason: the Wankel.

Fichtel & Sachs had enjoyed considerable success with its rotary Wankel engine in motorbikes, lawnmowers, pumps and snowmobiles. "It was obvious that this modern type of engine should also be applied in a power saw due to its vibration-free qualities and effortless starting," said company officials. F&S hoped that the acquisition of Dolmar would give them the foot-in-the-door into the tough international chainsaw market that without such a reputable name, would be a tough one to crack. The Wankel was indeed powerful, efficient, quieter than the standard chainsaw and easy to start. Like other Wankel-powered de-

Top: The 1949 CL, the first saw offered by Dolmar post-war. Well built with 247-cc JLO engine, but a dying breed. *Collector Marshall Trover, photo Brian Morris*

Above: The 1953 CP, a rugged gear-drive design influenced by the Hornet DJ 3500H. *Collector Marshall Trover, photo Brian Morris*

vices, its makers were proud of it and experts admired it, but it was so different from the prevailing technology that consumers were a bit wary of it. Although smooth-running and relatively quiet, the Wankel was heavier than regular saws of comparable size, and the engine tended to run hot. Unfortunately, what looked like a very worthwhile new direction did not turn out to be commercially viable. In 1991 Sachs-Dolmar was purchased by Makita of Japan, which now operates it as a subsidiary.

Above: The 1961 DB-50-2, electric bow was used primarily in construction and sawmills. *Collector Marshall Trover, photo Brian Morris*

Left: The 1957 Dolmar CF, a smaller, lighter (10 kg /22-lb) one-man saw with an all-position diaphragm carburetor. *Collector Marshall Trover, photo Brian Morris*

Below: The 1961 CC was the first direct-drive saw produced by Dolmar. *Collector Marshall Trover, photo Brian Morris*

Inset below: Dolmar sold thousands of this rugged little 1966 CA. *Collector Mike Acres, photo Vici Johnstone*

Right and bottom: 1975 Sachs-Dolmar KMS4 was the only chainsaw ever produced with a Wankel rotary engine. *Collector Mike Acres, photos Vici Johnstone*

Russian Chainsaws

During the 1920s, the Soviet government instituted a campaign to build up their heavy industry and military technology. The USSR's first chainsaws were electric. They became such a force in the nation's forestry practices that the dense forests of Siberia were strung with poles and electric cables. But the greater power and independence of gas-powered saws were the next step, and after Andreas Stihl made his initial sales triumph in 1931, the country undertook the production of its own gas saws.

The masthead of Russia's flagship chainsaw manufacturer translates into English as "The federal state unitary enterprise 'Machine Building Plant named after F.E. Dzerzhinsky.'" The company began as a shipbuilder in 1859 and began mass-producing saws in 1955. In 1958 its Druzhba gas saw won a gold medal at a Brussels trade show.

Today Dzerzhinsky remains a going concern. In an age where all chainsaws are starting to look alike, Dzerzhinsky's products can boast some genuinely distinctive features. For example, their biggest saws, the Ural 2-T Electron and the Druzhba 4-M Electron feature "high handles," an intriguing innovation that allows the operator to work upright, away from the saw's noise and exhaust. In a blast from the past, the cutting unit can be swivelled ninety degrees in order to allow the saw rather than the operator to adjust to different cutting positions.

For all that, they are not very big saws, with a maximum bar length of 18 inches (46 cm), although with the added swivelling mechanism and high handles they are heavier (about 26 pounds/12 kg) than comparable saws we have in the West. The Ural 2-T and the Druzhba 4-M can also be fitted with a "hydraulic wedge" to assist a single operator in felling large trees.

Dzerzhinsky's smaller models are the Tayga-245, at 20 pounds/9 kg and a 16-inch/40-cm bar lacking the high handle of the bigger models and looking much more like our western saws. The Ural-44 is even smaller at 15 pounds/7 kg and a 12-inch/31-cm bar. Dzerzinsky claims low noise and vibration for its saws (but who doesn't?) and aside from these features they seem to be quite like western saws with two-stroke, one-cylinder gas engine, centrifugal clutch, chain brake, etc.

Above and below: The high handlebars on this 1972 Druzhba may look strange but they are a sensible adaptation for falling close to the ground and bucking small logs without making the operator bend over. The unique but practical Druzhba was produced for decades with little or no change in design.

Collector Mike Acres, photos Lionel Trudel

Two views of the PPK Quick 80, a design derived from the Stihl KS43 that was produced in France to support the German war effort. *Collector Marshall Trover, photos Brian Morris*

French Chainsaws

France also burst into chainsaw production after the war. In the city of Courbevoie, Pierre Paul Klomp's Société PPK had produced all sorts of industrial cutting machinery, and in 1945 they came out with their first two-man chainsaw, weighing 60 kg (132 lb) with a 250-cc engine and a 1.5-m (59 in) bar. Its African colonies were important to the French economy, and PPK followed this first saw with more powerful models, with 350-cc and even 500-cc engines, especially for the African market. In Thiers in south-central France west of Lyon, René Sauzède's Société Rexo put out the two-stroke Model RS after the war, and the four-stroke Model RI, both weighing around 60 kg. A smaller outfit in Paris, la Société Vamos made both electric and gas-powered saws, intended for use in mines and sawmills as well as for forest work.

Norway and Jo-Bu

The Norwegian logging industry faced problems similar to those on British Columbia's coast. Just as the terrain of the steep, rocky BC inlets made big, heavy saws difficult to take into many parts of the woods, the rugged countryside lining Norway's fjords presented similar obstacles. The handsaw was still much easier to move around, as well as cheaper and more practical.

Gunnar Busk, an experienced gunsmith, conspired with Trygve Johnson and Anders Skuterud to develop the first Norwegian-made chainsaw—one that would suit local conditions better than imported saws.

The partners designed a one-man prototype saw and Johnson and Busk combined their surnames to create a new company, Jo-Bu A/S Mekaniske Verksted. They established a factory in Drøbak, south of Oslo, and in 1948 put their prototype into production as the Jo-Bu Senior.

The Senior was powered by a 125-cc Aspin engine and with an 18-inch (45-cm) bar it weighed 38.5 pounds (17.5 kg). Its float carburetor could be easily rotated for felling, angle cuts or bucking, and it used a scratcher chain in the "straddle" style—instead of a single drive link running in a groove, the bar had no groove and the chain was propelled by two drive links that straddled the bar.

The Senior was an immediate success, and over the next few years Jo-Bu sold 7,300 of them. In 1952 they introduced a much lighter saw, the Jo-Bu Junior. By now, Jo-Bu was making its own 76-cc engine, which it combined with a lever-controlled clutch and its proven rotating carburetor, straddle bar and chain design. Made of sand-cast aluminum, the Junior weighed 23 pounds (10.4 kg) with an 18-inch (45-cm) bar. It was a very lightweight saw for its time and sales took off—the company eventually sold almost forty thousand Juniors.

In 1957, Jo-Bu introduced the "Viking," which was very similar to the Junior and used the same engine. In 1958 it brought out a larger saw, the Jo-Bu Model 93, which suffered from an irreparable machining flaw and had to be dumped into the deepest part of the Oslofjorden.

The 1952 Junior, the second saw model produced by Jo-Bu, was ideal for use in Scandinavian forest work and sold spectacularly. It had straddle link chain, a manual clutch, automatic rewind starter and the rear handle swivelled to allow tree falling.

Collector Marshall Trover, photo Brian Morris

The 1964 Tiger, a 93-cc direct drive and second most popular Jo-Bu model. *Collector Marshall Trover, photo Brian Morris*

They corrected the flaw in their 93-cc engine, incorporated it within their successful Junior design and in 1959 announced the Jo-Bu Junior Super. This saw paved the way for the company's next big success, the 1960 D94 Tiger. Weighing only 10.5 kg with its 93-cc motor and all-position Tillotson diaphragm carburetor, the Tiger was offered in a choice of colours: black, green or burgundy, all with red trim. It was a commercial success, selling over forty thousand units, and was the vehicle for Jo-Bu's transition from straddle bar and chain (sold with the early Tigers) to the prevailing style of single-drive link chain running in a grooved guide bar (which was offered on the later models of Tiger).

In 1964 Jo-Bu followed the success of the Tiger with the more powerful Tiger S, and then hit a snag with the 69-cc (4.2-cubic-inch) Starlet, which had a new starter design that tended to clog with sawdust. Jo-Bu did somewhat better with its next saws, even making some sales to the large USA market in 1967 with a saw the same size and weight as the Starlet, the M5. It was the age of teensy consumer saws, and the next year Jo-Bu did even better with the 56-cc (3.4-cubic-inch) Model L6 at only 15.5 pounds (7 kg) with a 14-inch (35-cm) bar. They were gradually making inroads into the US market, not setting up a proper distributorship there until 1972, but meanwhile they managed to sell over forty-six thousand of the L6. The 1970 Model L7 introduced the first anti-vibration mounts on a Jo-Bu saw, and throughout the 1970s the company did well. They began to include a new development, the chain brake, on some of their saws, and had success with such models as the LP4, which with a 48-cc engine weighed only 5.9 kg (13 lb).

The LP series was popular because among other things, the designers had managed to significantly reduce vibration by using AV mounts, rebalancing the crankshaft and using a technique they called "handle tuning." In 1976 Jo-Bu struck an agreement with the Swedish company Partner, and the Partner P48 and P49 saws were actually versions of Jo-Bu LP-model saws.

However, Electrolux was busily buying up chainsaw manufacturers, and in 1978 the huge conglomerate bought Jo-Bu along with its fellow Scandinavian companies Jonsered, Husqvarna and Partner. The very last Jo-Bu saw appeared in 1982: Model 949, which was actually a Jonsered 490 with a different label. Today, Electrolux still uses the Drøbak plant to make guide bars and other parts for their Jonsered and Husqvarna saws.

The Diesel Chainsaw: Rasmus Wiig's "Comet"

Because of their high combustion temperatures and cylinder pressure, diesel engines are more heavily built than gas engines. Most of us associate them with trucks and buses, trains and ships, or with small German cars driven by enthusiasts who would never go near a gasoline-run vehicle. But not with lightweight outdoor power equipment.

In fact, the first diesel chainsaw, the "Comet," was made in Oslo, Norway in 1949 by an ex-mine captain named Rasmus Wiig. At 19 pounds (8.5 kg) it was, by the standards of the time, sensationally lightweight, being 6 pounds lighter than the landmark McCulloch 3-25, which came out in the same year. Diesels take a bit more effort in starting than gas engines. Many use an electrically heated "glow plug" to begin the dieseling process but instead of an electric element Wiig used an ignition plug, or *tändror*.

Like the glow plug in other diesel engines, the tändror quickly became red-hot during the Comet's starting sequence. However, the heat source was not an electric element, but propane gas which was stored in the saw's tubular handlebar. Just like firing up a propane torch, the operator would open the valve and ignite the flame that would quickly heat the tändror. When it began to glow, a pull on the starter rope would start the engine, and the operator would turn off the propane. The handlebar was said to hold enough propane to start the saw several hundred times. Wiig claimed that the Comet was so easy to start that, should it ever stall once it was warmed up, the operator could simply spin the pulley by hand in order for it to kick back into life.

Wiig was certain that his fortune was secured. From his small Oslo factory, "Norsk Sag-bladd Fabrikk" (Norwegian Saw Blade Factory), he began turning out the Comet Model A. He certainly had good reason to be optimistic. His two-cycle engine, like other diesels, used a fuel injector rather than a carburetor, meaning the motor ran fine at any angle just like the hot new McCulloch. It would run on either diesel fuel, kerosene, or a gasoline/oil mix. Damp and even torrential conditions, Wiig boasted, wouldn't slow the path of this Comet. One of his ads showed him dowsing a Comet in a barrel of water, pulling it out, lighting up the tändror and starting the saw immediately—unthinkable for a gas-powered saw with electric ignition.

Cold weather did present a problem, however, and in Norway there is a lot of that. In winter the propane in the handlebars could lose pressure. When this happened, the operator would be called upon to remove his mittens and warm the bar with his hands,

Rasmus Wiig's semi-diesel chainsaw was surprisingly light and durable. It would run upside down and you could start it after dousing it in a barrel of water.

189

COMET

something even a Norwegian lumberjack could find tiresome if repeated too often at sub-zero temperatures. Despite the good impression the Comet made on everyone who saw it in action, sales remained in the hundreds rather than the thousands Wiig had hoped for. Wiig licensed the Comet to M.T. Bjerke Company's "Como" operations, hoping the larger firm would provide the market reach he was unable to. This gave birth to a new Comet—the Swedish Model B, which sold approximately one thousand units between 1950 and 1953. This disappointing result did not come from lack of trying—in 1953 M.T. Bjerke even introduced a new model, the Comet S with an automatic rewind starter mechanism called "Magnapull."

In 1954, M.T. Bjerke and Como gave up on the Comet—but not on the diesel chainsaw. Just as they phased out the Comet, they contracted with the Swedish manufacturer Jonsered to produce chainsaws. Since Como's expertise depended so strongly on what they had learned from Rasmus Wiig's design, the first Jonsered model was a propane-heated diesel saw, the P (for "propane"). Not judged to possess the commercial potential that was needed, the P was never marketed, being instead immediately supplanted by a similar diesel model, the XA "Raket," a saw notable for having to be started upside down.

Jonsered produced several models of diesel saws, including the XA and the similar XA-19. For the overseas market, they also produced a very limited edition (possibly no more than one hundred) of the XC, a diesel saw in which the tändror was replaced by the more common electric glow plug powered by two "D" cell batteries. In the chainsaw boom of the 1950s, diesel saws made a brave start, but despite their seeming advantages soon fell behind and vanished.

The 1954 Comet S semi-diesel. *Collector Mike Acres, photo Lionel Trudel*

To start the 1954 Comet S semi-diesel you had to heat a glow plug using propane stored in the handlebar. *Collector Mike Acres photo Lionel Trudel*

Top: The 1955 Jonsereds XC was a reprise of the Wiig-Como semi diesel that was started with an electric glow plug powered by flashlight batteries. *Collector Marshall Trover, photo Brian Morris*

Below: The 85cc XD, produced in 1958, was Jonsereds' second gas-powered saw. *Collector Marshall Trover, photo Brian Morris*

Jonsereds

Jonsereds

Jonsered is the name of the town near Gothenburg where the Jonsereds Fabrikers AB factory was located originally. Jonsereds (the "s" was dropped later) started in 1833 in the textile industry, later expanding into the forest industry. In 1950 Jonsereds was asked to make parts for the Comet diesel chainsaw; it was not long before it had bought Comet and started a new brand, Raket (Swedish for rocket).

The first gasoline-powered Jonsereds model was the 1957 XB but it was underpowered so in 1958 the XD replaced it, having an 85-cc (5.1-cubic-inch) engine. The gear drive was virtually obsolete for Scandinavian woodcutting but the XD Super with a higher gear ratio was offered and remained in production until sales dropped altogether.

The first direct-drive Jonsereds was the 1960 XF. With a 110-cc (6.7-cubic-inch) power-plant, the XF enjoyed good sales through 1969. In 1963 the XG and XH were introduced, both with 85-cc (5.1-cubic-inch) displacement. The first saw with the classic Jonsereds design and shape was the 60, which appeared in 1966. It had a 56-cc (3.4-cubic-inch) engine and essentially started Jonsereds on a market expansion. Other models followed: the

62 with anti-vibration handles, then the 75, 80, 50, 52, 45, 90 and 110. There were more derivatives of these models until Electrolux purchased Jonsereds in 1979.

Under Electrolux, design tended to be by family—with Jonsered, Husqvarna and Partner models all sharing common parts—though Jonsered stuck with its 87-cc (5.3-cubic-inch) model until the early 1990s when the new 94-cc (5.7-cubic-inch) Electrolux series was produced and became the Jonsered 2094.

Above: This 1976 49cc Model 49SP was a well-liked pulp cutting machine of classic Jonsereds design. *Collector Mike Acres, photo Vici Johnstone*

Left: The 85-cc XH was released in 1963 and produced until 1968. *Collector Mike Acres, photo Vici Johnstone*

Below: The 1960 XF was a 110-cc workhorse. *Collector Mike Acres, photo Vici Johnstone*

PARTNER®

PARTNER
R17

MED 4✚ FÖR EGEN SÄKERHET

BE-BO and Partner

The first chainsaw manufactured in Sweden came in 1949 from the AB Bergborrmaskiner company, which had been in the business of building rock drills. The saw was called the Be-Bo. The success of this model led the company to set up a new factory in Molndal and develop a new model, the C6. A competition was held among the company employees to choose a name for the new saw and the winner was Partner. The company became AB Partner.

The C6 had a whole range of attachments in addition to the chainsaw transmission, bar and chain. There was a clearing saw attachment, an outboard motor attachment, a cultivator attachment, an earth drill attachment, an ice drill attachment, a water pump, a winch and others. The C6 was not ideal for the small log work done in Scandinavia so in 1958 Partner introduced Sweden's first direct drive, the R11, which had a 90-cc engine that gave it considerable cutting power. Also at this time, an X21 model with a Gilmer Belt reduction drive system appeared for export sales only.

The one drawback of the R11 was the straight stub-type rear handle, but this was rectified in 1962 with the release of the R12 model, which had a pistol-grip rear handle. The engine of the R12 was used to power the K12, the first gas-driven metal-cutter: it was not long before most fire departments around the world had a new rescue tool on their trucks, for years referred to simply as the "K12." In 1962 Partner added the model TS, a 70-cc saw manufactured by Nymanbolagen AB (Crescent), specialized for limbing in small log work (Husqvarna sold the same saw as their model E70 and then F70). The next model brought out by Partner was the R14 with a 76-cc engine. Then in 1967 they introduced the R16, a completely new design with a 55-cc high-speed engine weighing in at 13 pounds (5.9 kg) for the powerhead. In Sweden it was usually equipped with a 13-inch (33-cm) bar and made an ideal pulpwood saw. A year later Partner added the R17, the first Partner saw with anti-vibration handlebars, to its line. A year after that they introduced the R18, the first saw with heated handles.

In 1971 Partner introduced the 55-cc R17T, with transistorized ignition and heated handles, plus their first 65-cc saw, the R20, designed mainly for export because bar lengths in Scandinavia were usually no longer than 13 inches (33 cm). Partner also brought out a model with a larger 85-cc engine, the R30.

In 1972 the 100-cc R40 made its debut. In 1973, Partner produced its first saw with a chain brake, the R22. In 1975 it improved on the R22 with the P50, which had an inertia-activated chain brake. Among their many other models was the F55, a 55-cc "Farmer" saw intended for the non-professional user. In 1978, Partner introduced a new design of saw with Jo-Bu of Norway, the 48-cc P48.

Electrolux purchased Partner in 1979 and carried on with Partner-designed models, starting with the completely new 50-cc Model 5000. Each year there were improvements and model changes in the 55-cc, 65-cc, 85-cc and 100-cc machines that stayed on the market until well into the 1980s The engine that started out as the R20 in 1971 was still used in Partner power cutters in 2006. Skil and Frontier built small saws for Partner, but in 1988 the marketing of chainsaws with the Partner name in North America was replaced with the Poulan Pro brand name. All in all, there had been at least eighty-four Partner saw models from the C6 through to the 1980s, many still running today.

The 1949 Be-Bo, the first chainsaw made in Sweden, was modelled closely on the Canadian-built Hornet DJ-3500H.
Collector Marshall Trover, photos Brian Morris

The Be-Bo had a 125-cc engine, weighed 9 kg (20 lb) fuelled up, and did so well the company changed its name to Partner and became a major force in the chainsaw industry.
Collector Marshall Trover, photo Brian Morris

Top: The 1955 C6 was the first model to bear the "Partner" name. *Collector Marshall Trover, photo Brian Morris*

Inset above left: The 1958 R11, the first direct-drive chainsaw manufactured in Sweden. The improved 1962 R12 version (inset right) had pistol-grip controls. *Collector Mike Acres, photo Lionel Trudel*

Right: The P85 powered a metal cutter famous in rescue work. *Collector Mike Acres, photo Vici Johnstone*

HUSQVARNA

Early Husqvarna factory. *Courtesy of Husqvarna*

Husqvarna

One of the newest players on the international chainsaw scene is actually by far the oldest company. Husqvarna of Sweden dates its founding to 1689, and evidently can trace its roots as a business even farther back than that. The reason for an industry's precociousness, true now as much as it was then, is access to energy sources, and by the seventeenth century Sweden was a world capital of the era's cutting-edge energy technology: the water wheel. In a mountainous northern land with many lakes, rivers and streams, the Swedes had been using water wheels since the middle ages—first to grind grain, and then for other applications such as sawing wood and fulling (cleaning and thickening) cloth.

By 1689, the castle Rumlaborg had stood on the Humblaruma River for several centuries, in hilly country about 150 miles (240 kms) southwest of Stockholm. Here, the force of the water flowing out of Lake Vättern gave its mill a perpetual source of power. It was the estate's mill, otherwise known as the "house mill." The Swedish word for house is *hus*, and mill is *kvarna*, and by the sixteenth century the river was known as the Husqvarna River. As part of a military buildup in 1689, King Charles XI established an arms factory at Husqvarna that was soon producing 240 musket bores a week.

As Sweden's relationships with its European neighbours improved, the government de-escalated arms production, and eventually the Husqvarna works was auctioned off to private interests. By the end of the nineteenth century the firm was famous for its sewing machines, and in the twentieth century began manufacturing kitchenware, grinders, woodstoves, bicycles, motorcycles, mopeds and refrigerators. In the 1950s it even pioneered home microwave ovens. But it still hadn't tackled the signature product with which the firm is now so much identified.

By the time Husqvarna entered the chainsaw market, there were already four other chainsaw manufacturers in Sweden alone. Husqvarna's lateness may have been an example of corporate inertia rather than lack of individual initiative in the company itself. During the 1930s its head designer, Folke Mannerstedt, had pioneered a successful line of motorcycles using light alloys, and in the 1940s and '50s his successor, and later managing director, Calle Heimdahl continued his success with the Svartkvarnas motorcycle.

Heimdahl was an inventive designer who also experimented with motorized saws. Using a motorcycle engine, he built a reciprocating drag saw, and with a 50-cc moped engine he built a prototype chainsaw. But company management at the time had no great interest in chainsaws. However, Husqvarna owned a sawmill at Taberg, just a few miles downriver from its main works at Jönkoping. The mill was run by an experienced lumberman, Gösta Arneklo, who was convinced that chainsaws were a growth industry and that Husqvarna, with its huge facilities for research, development and production, was missing the boat.

Finally in the late 1950s, the corporate giant started to move. In 1959 Husqvarna premiered the one-man Model A90 saw. At 90 cc and weighing just over 25 pounds (11.5 kg), despite a few idiosyncrasies it was a well-built saw; it ran about 10 percent more quietly than saws of comparable size, in an era when ear protectors were not yet standard equipment. The A90 sold moderately well, about 350 units to the professional market in its first year, just enough for Husqvarna to consider itself in the chainsaw business.

In 1962, along with its purchase of a firm called Monark-Crescent, Husqvarna acquired Monark's design for a chainsaw built around an outboard motor engine. This saw appeared under three different names: Crescent issued it as the 06, it was offered by Partner as the TS, and by Husqvarna as the 70, a saw light enough to be marketed as "a saw for trimming"—introducing the idea that not only was the chainsaw now a one-person tool, it was now light enough to be lifted up for limbing work.

Once established in the chainsaw market, Husqvarna had the resources, including the design talent, to keep up with the front-runners in saw design. In 1969, for example, it introduced the Model 180, one of the first saws in which anti-vibration mounts were integrated into the design. Part of Husqvarna's success during these years came from it being willing and able to buy up the competition, as it did with Monark and in 1970, the clearing-saw company Tandsbyns Mekaniska Verkstad. But as sure as big fish eat little fish, the company in turn was purchased by Electrolux in 1977. The electric-motor firm must have felt it was onto a good thing, because after taking over Husqvarna it soon bought Jonsered, Partner, Poulan, Pioneer and Jo-Bu.

Soon the chainsaw world, once inhabited by hundreds of small companies, was pretty much divided up between Stihl and the Electrolux/Husqvarna conglomerate, with a few smaller concerns doing their best to preserve some share of the market. Thirty years after Husqvarna sold their first saw, the head of their chainsaw division summed it up when he said, "There's only the odd country here and there across the world where you won't find our saws."

Although Husqvarna didn't enter the chainsaw business until 1959, their experience in marketing and manufacturing helped them make up for lost time. Here champion motorcycle racer Bengt Håkansson demonstrates an early A90 to a Swedish logger. *Courtesy of Husqvarna*

Right: The 1975 Husqvarna 2100 was an extremely popular 99-cc professional model capable of handling up to a 5-ft bar.
Collector Mike Acres, photo Vici Johnstone

Below: The A90, the historic first chainsaw produced by Husqvarna in 1959, had a 90-cc engine, direct drive, weighed 5.25 lb (11.5 kg) and was very quiet—all in all a good start.
Collector Mike Acres, photo Lionel Trudel

These saws were considered to be the first Swedish built direct-drive saws safe for limbing. Above 1963 Partner TS and below, 1964 Husqvarna F70. *Collector Mike Acres, photo Lionel Trudel*

The 1954 BLK, the first practical one-man saw from Stihl, was produced in large numbers for the European market and the German Army. *Collector Marshall Trover, photo Brian Morris*

STIHL®

Stihl Comes Back

Through the showing of your original machine in Canada and the United States, you taught the [logging] industry the possibilities of using gasoline engines for cutting down trees as well as bucking them into log lengths. Throughout the United States and Canada, you were the first one to demonstrate a practical, lightweight gasoline-engine [chain] saw for this application. Today there are at least 100,000 chain saws being used in Canada and in the United States for this work, all of which are more or less copies of your original machine.

—*Arthur Mall, letter to Andreas Stihl, 1948.*

From the bombed-out Bad Canstatt machine works, Andreas Stihl Maschinenfabrik salvaged what it could and moved it to a new site in Waiblingen-Neustadt, a village on the Rems River about eight miles east of Stuttgart. Reconstruction was slow, but with the end of hostilities, the international market was accessible once again, and Stihl resumed exporting saws in 1947.

But for at least two years, the company did business without the guiding hand of its CEO. Although Andreas Stihl had taken part in the war effort under some duress—"You could either get sent to a concentration camp," he later said, "or you could do business with them"—he was kept out of circulation, and for a short time the new works at Waiblingen were put on a "disassembly list."

In 1948, however, with Andreas back at the helm, the company began working on new models, and in 1950 they came out with their first one-man chainsaw, the BL. The BL weighed 16 kg, was designed to take up to a 32-inch (80-cm) bar and just to be on the safe side, still had a removable head end mount to accommodate a second operator.

Although all chainsaw manufacturers were coming out with one-man saws, Stihl managed to remain a serious contender. In 1942 the company had devised an "ice saw," a slide-mounted model with a roller-nose guide bar to facilitate non-stick cutting of blocks of ice (still commonly used for refrigeration), or to cut through the surfaces of frozen lakes for fishing. In 1950 they introduced the friction-reducing roller nose to their regular guide bars. Stihl had started to rebuild its domestic market with such saws as the RBB 50 bow saw in 1952 and the 1953 one-man 15-kg LP. In 1954 the company scored a hit with the BLK, which at 24 pounds (11 kg) and a maximum 24-inch (60-cm) bar, had no helper-handle option. Also in the 1950s Stihl introduced two electric saws that became very popular: the direct-drive ESL and later the EL, useful not only to backyard woodcutters but to millers, joiners and carpenters.

In 1959 Stihl introduced the Contra, their first one-man direct-drive saw. At 26.4 pounds (12.2 kg) with a 17-inch (43-cm) bar and chain, all the modern conveniences such as an automatic chain oiler, super cutting power and durable German engineering, the Contra was a real contender. Sold in North America under the name "Lightning," it gave Stihl a dramatic rebirth in that wealthy and fast-growing overseas market. It was soon followed by a model that featured anti-vibration mounts (see sidebar "White Finger," page 206), that are now considered essential to any chainsaw. In 1971 Stihl introduced a chain brake that, in the case of kickback, declutched the engine as well as stopping the chain. New models poured forth, each one setting new standards for design and performance.

With breathtaking speed, Stihl regained the upper hand in chainsaw development that it had surrendered before the war, opening an American factory in Virginia Beach, VA in 1974, starting out with its first US-produced saw, the 015. Weighing 8 pounds (3.6 kg), the 015 was a lightweight model with the top handle "arborist's saw" design that enables one-handed use. Adding one trend-setting model after another, by the 1980s Stihl could well justify its claim to be the world's largest chainsaw manufacturer, its place in the hierarchy threatened only by the advance of Husqvarna/Electrolux.

Above and below left: The 98-cc, 24-lb (11-kg) BLK was imported and sold in the US in limited numbers but could not compete with the US brands of the time. *Collector Marshall Trover, photo Brian Morris*

The saw that restored an empire. Released in 1959 as the Contra in Europe and in North America as the Lightning, it was Stihl's first direct-drive saw and proved a workhorse for larger timber cutting. It re-established Stihl as a leader in saw design and repositioned the company for a return to dominance in world markets. *Collector Marshall Trover, photo Brian Morris*

Left: The Stihl Lighting. *Collector Marshall Trover,*
photo Brian Morris

**Below: Introduced in 1969, the 090G was an improved version
of the Contra G with the 106 cc (6.5 cu. in.) engine and 2:1
reduction transmission. A big timber workhorse, it could be
used with guide bars as long as 150 cm (60 inches).**

Collector Marshall Trover, photo Brian Morris

The Twenty-First Century

Mike Acres (left) and Art Patterson get a Remington Bantam ready to have its portrait done. *Photo Vici Johnstone*

Throughout the late twentieth century, a host of manufacturers thrived. The two-man chainsaw disappeared, and the lightweight casual-user chainsaw market exploded. As the century progressed, chainsaws got smaller and smaller, and certain features became common. Plastic housings, if not as strong as metal, were lightweight, surprisingly durable, and cheap to replace. The handle and throttle trigger, the choke on the right and the starter rewind and shutoff switch on the left, became industry standards. It began to get harder to see the differences between saws from different manufacturers.

By the 1980s, safety features such as anti-vibration mounts, to soften the effects of extended vibration on a user's hands, were becoming standard on saws. So was the chain brake—first contact-operated, so that when kickback jerked the bar upwards, the brake handle hit the operator's left wrist and the chain stopped moving—then inertial, so that the jerking motion itself stopped the chain. The operators themselves had more safety options than ever: not only hard hats but ear protectors, plastic goggles, rubberized gloves, Kevlar chaps and pants and steel-toed boots.

White Finger

Describing the test runs of the first Dolmar chainsaws in Quebec in 1937, a Canadian Pulp and Paper Association writer mused, "The vibration of the handles on the motor end of the saw made it necessary to change men on this end every 20 minutes. The men would likely get used to this vibration in time but due to the weight of the saw when used steadily it would be just as well to have the men change ends." The workers did get used to the relatively low-frequency vibration of those original gear-driven saws, but the real trouble began when the high-revving motors of later models like the Stihl Lightning introduced a whole new order of high-frequency engine vibration. The result of spending all day, day after day, handling the screaming two-stroke engines was a horrific condition called "white finger."

Chainsaw History: A Timeline compiled by Mike Acres

1858   First patent obtained for a saw chain

1918   Shand obtains patent on saw chain (Canada)

1919   Sector saw introduced in Sweden

1919   Gerber produces portable bandsaw

1920   Wolf electric chainsaw introduced

1925   Rinco introduces first gasoline chainsaw

1926   Stihl and Dolmar introduce electric chainsaws

1927   Stihl and Dolmar produce gasoline chainsaws

1927   Wolf introduces first pneumatic saw

1939   D.J. Smith Equipment Co. produces its first gasoline saw, the Model A

1940   Atkins introduces models X and Z, heavy-duty professional electric chainsaws

1940   Danarm produces Mark 1, two-man saw for British army

1940   Hassler improvements of saw chain patented, including bent-over teeth

1942   Disston begins producing its first saw, the G-10, for the US Army

1943   Lombard introduces its first gasoline, electric and pneumatic saws

1944   IEL introduces world's first one-man chainsaw, the Beaver

1945   Mall introduces its electric chainsaw, the Universal

1946   Poulan builds his first chainsaw, using Homelite-made engine

1947   Joe Cox introduces Oregon Chipper Chain

1947   Homelite introduces electric chainsaw model ECS with matching generator

1947   Sears introduces Craftsman 351.23001 saw, built by Reed-Prentice

1947   PM introduces first chainsaw model, the Universal

1947   Danarm Canada introduces 98 Hi-Baller one-man saw and 250 Hi-Baller two-man saw

1948   Disston introduces one-man saw, model DO-100

1948   McCulloch introduces its first chainsaw model, the 1225A

1948   Jo-Bu introduces its first saw model, the Senior

1948   Armstrong Products begins manufacture of chainsaw clutch; within one year is world's largest manufacturer of chainsaw clutches

1948   Hornet introduces model DJ-3500H, its first one-man chainsaw model

1949   McCulloch introduces its first one-man chainsaw, the 3-25, the first US-built all die-cast one-man saw

1949   Burnett introduces its last saw model, the Powermatic

1949   Dolmar introduces models CB-35 and CB-50 and CB65, one-man saws with bow-type cutting attachments

1949   Homelite introduces first saw, Model 20MCS, 30 lbs, utilizing Gilmer belt reduction

1949   Dolmar introduces model CL, all die-cast two-man saw with 247-cc ILO engine

1949   PM introduces Woodboss version of its first one-man chainsaw

Teles Smith SW4. Collector Marshall Trover, photo Brian Morris

Whitehead, Model 40. Collector Marshall Trover, photo Brian Morris

NSU, Ural. Collector Duane Zollo, photo Brian Morris

Talkie Tooter. Collector Marshall Trover, photo Brian Morris

Echo CST 610EVL Twin. Collector Marshall Trover, photo Brian Morris

Spear and Jackson, Challenger. Collector Marshall Trover, photo Brian Morris

1949	First chainsaw manufactured in Sweden is introduced by Bergborrmaskiner AB, designated the Be-Bo
1950	Stihl introduces model BL
1950	PM introduces Redhead one-man/two-man saw model
1950	First Walbro carburetor sold to Clinton
1950	Armstrong introduces first model of its famous rewind starter, later bought out by Fairbanks Morse
1951	IEL introduces world's first direct-drive chainsaw
1952	Reed-Prentice introduces Timberhog Bantam model with automatic chain oiler
1952	Shindaiwa Kogyo Co., founded in Japan, manufactures electric chainsaws
1952	*Chain Saw Age* magazine publishes first issue in August
1952	PM Introduces the Torpedo one-man/two-man saw model
1952	Jo-Bu introduces the Junior
1952	Nielson chain grinder introduced
1953	Tillotson introduces first model of diaphragm carburetor Type H for gravity feed and Type HP with fuel pump
1953	Poulan introduces its Model A, the first one-man saw complete with bow configuration
1953	Beaver Saw Chain Co. established in Portland, Oregon by Harry, Art and Arnold Siverson
1953	Dolmar introduces one-man saw model CP
1953	PM introduces Rocket K1 model with belt drive reduction
1953	Mall now selling chrome plated "planer" and "scratcher" chain for replacement market
1953	Strunk introduces first chainsaw model
1953	Lebanon Metal Products introduces model 200
1954	Druzhba introduces first chainsaw model
1954	PM introduces Rocket K3 with new Tillotson HP all-position carburetor
1954	Homelite introduces the model 17 lightweight gear-drive saw
1954	Jonsered introduces model XA diesel chainsaw
1955	Monark introduces Silver King model, direct drive with 3 hp
1955	Partner introduces model C6, one-man gear-drive saw
1955	Indian Chainsaws appears with Model 351-D, manufactured by Luther Corp of Rockford Illinois
1955	Cobra chainsaws enter market with low cost direct-drive
1955	Wasp chainsaws appear for first time
1955	Titan brings out Model 70 and lightweight gear-drive Model 35
1955	Tree Farmer Chainsaws introduces model AL-P, 4.7 cu. in. with all-position carburetor
1955	Mall introduces "Double Cutter" planer chain
1955	Homelite introduces very successful model 5-20 lightweight gear-drive saw
1955	McCulloch introduces model Super 33, and its first direct-drive saw, model D33
1955	H & S Chainsaw appears for first time

1955	Fate-Root-Heath Co. introduces saw chain with insertable cutters
1955	England Motors enters chainsaw market, located in Pine Bluffs, Arkansas
1956	Homelite introduces its first direct-drive saw, model EZ
1956	Remington Arms Company purchases Mall Tool Company
1956	Outboard Marine Corp. purchases IEL
1956	Sensation Lawn Mower Co. of Ralston, Nebraska enters chainsaw market
1956	Lancaster introduces "Nifty" chainsaw, selling for less than $150
1956	Gouger Saw Chain Co. begins production of 1/2-inch pitch saw chain in Hamilton, Ontario
1956	McCulloch introduces Pintail saw chain
1956	Porter Cable Co. of Syracuse, NY introduces Model 530 chainsaw
1956	Atkins introduces 700 series chainsaw chain
1956	Hoffco introduces two new gear-drive saws, one direct-drive
1956	Monark Silver King introduces two models designed to allow "one-hand" operation
1956	Magnaflux of Chicago forms division to manufacture Indian Chainsaws, formerly made by Luther
1956	Trams Chainsaw Co. of Chicago introduces new model with price starting at $169
1956	Kiekhaefer begins marketing model KA-211
1957	Dolmar introduces model CF
1957	Jo-Bu introduces the Viking chainsaw
1957	PM introduces Model 21 direct-drive with 82-cc engine, aka "The Canadien"
1957	Jonsered introduces its first gasoline chainsaw, the XB (62 cc)
1959	Stihl introduces one-man direct-drive chainsaw "Contra" in Europe, "Lightning" in North America
1959	Husqvarna introduces Model A90, its first chainsaw
1960	Textron acquires Terry Machinery of Montreal, distributor for Homelite in Canada
1960	Dolmar introduces the gear-drive CF (19 lbs.) to US market
1960	PM introduces Canadian model 270, a 5.8-cu.-in. (95-cc) lightweight saw
1961	McCulloch is first manufacturer to advertise chainsaws on network television
1962	Husqvarna introduces Model A100
1963	Carlton Saw Chain Company formed in Milwaukie, Oregon by Ray Carlton
1963	McCulloch introduces only chainsaw ever built with balanced piston engine, later withdrawn from market because of safety concerns due to over-revving
1968	Dolmar introduces famous Model CT (118-cc, direct-drive)
1969	Stihl introduces 090G model (106-cc, gear-drive)
1982	Alpina purchases gasoline chainsaw division of DESA Industries (Remington)
1982	Husqvarna introduces Rancher 50 and 266
1982	Oregon introduces Super Guard type 72LG, 73LG chain
1982	Townsend Saw Chain introduces Sabre Tri-Raker anti-kickback chain

Von Ruden. Collector Marshall Trover, photo Brian Morris

Sankey Aspin, Mark 2. Collector Marshall Trover, photo Brian Morris

Wright reciprocating saw. Collector Marshall Trover, photo Brian Morris

Mike Acres

Marshall Trover

The Chain Saw Collectors Corner

The Chain Saw Collectors Corner is an online archive, the world centre for chainsaw history and information. It is run by Mike Acres of Burnaby, British Columbia, at:

http://www.acresinternet.com/cscc.nsf

In writing this book I have consulted many sources, but Mike Acres' extensive personal knowledge, and the mountain of data he has assembled for display on the CSCC, have both been invaluable.

Mike's lifelong love of the saws began at an early age, when he began working at the McCulloch dealer in his hometown of Grand Forks, BC. Over the years he worked in many aspects of the forest industry, including selling McCulloch, Remington and Pioneer chainsaws.

Mike established the Chain Saw Collectors Corner in 2000. We ask anyone who reads this book and would like to offer new information on the history of chainsaws to send it to chsawcollector@gmail.com. Revisions or corrections should be sent to the author c/o Harbour Publishing.

Marshall Trover

Marshall Trover was born in Renton, Washington, in 1948 and grew up in the small mining town of Black Diamond. From as young as age eight, Marshall helped his father Orville cut, split and haul firewood to heat their home. They used obsolete two-man chainsaws to cut up old-growth fir logs that had been left behind by past logging operations. They started with a two-man Mall 6 and later a Titan 75 (its picture is in the book). Marshall's Uncle Ira, a saw mechanic at a Titan dealership, was the source of these saws and an excellent teacher of saw repair. Marshall spent one summer setting chokers at a logging operation for Weyerhaeuser but has spent the majority of his working life as a maintenance manager at the Boeing Renton 737 plant. Marshall's hobbies of collecting and restoring antique chainsaws and collecting hand-logging tools and chainsaw and logging literature has resulted in one of the most impressive collections in the world. He owns over 500 saws: many of them pre-1960, 150 of them two-man saws, and many of them low-production machines from the early days of chainsaw development in the 1930s and '40s. He can be contacted at marshalljtrover@msn.com.

Select Bibliography

Books, publications and major articles consulted

Acres, Mike. *Arthur W. Mall bio*, unpublished, May 22, 2006.

_____ *Power Machinery Ltd.*, unpublished, July 11, 2006.

Addison, J.D. and Jack Challenger. "History of the Power Saw," *Loggers' Handbook*.

Challenger, Jack W. "Power Saws: Experiments in British Columbia," *The Timberman*, August 1937.

Drushka, Ken. *Working In The Woods* (Madeira Park, BC, 1992).

Hall, Walter. *Barnacle Parp's Chain Saw Guide*. Emmaus, PA: Rodale Press, 1977.

Husqvarna AB, *300 Years and Aiming at the Future: A book on Husqvarna*, Sweden, 1989.

Knight, B.P. (Peter). *Down! In Three Minutes: A history of DANARM Chain Saws 1941–1984*. Boondall, Australia: B.P. Knight, 1998.

_____ *A Short History of Stihl Chainsaws*, self-published newsletter, date unknown.

Morrison, Ken. "Chain Saw Development—The Definitive History," *Power Equipment Trade*, January 1995.

Seventy-Five Revolutionary Years. Waiblingen, Germany: Andreas Stihl AG & Co., 2001.

Solo Post 1948-1998: Commemorative Publication on the Occasion of SOLO's 50th Anniversary on 10th February, 1998. Sindelfingen, Germany 1998.

Wardrop, Jim. "British Columbia's Experience With Early Chain Saws," *Material History Bulletin, History Division Paper No. 21* (Ottawa: National Museums of Canada, 1977).

Weaver, Buck. "Idaho Power Saw Tests," *The Timberman*, page & issue unknown, 1935, quoted in *Chain Saw Age*, September 1986, p. 65.

Periodicals consulted

British Columbia Lumberman, Chain Saw Age, Loggers' Handbook, Sawdust From The Kerf, Southern Lumberman, The Timberman, The Truck Logger, Timber of Canada, West Coast Lumberman

Collections

Mike Acres collection, Burnaby BC

Marshall Trover Collection, Renton WA

Interviews

Acres, Mike; Challenger, Dave; Trover, Marshall; White, Frank.

Index

Acknowledgements

After working on this book for a year and a half, the project really started to blossom when I went to visit the Chain Saw Collectors Corner headquarters in Burnaby and enjoyed the generosity that Mike and Judy Acres extended to me both at work and at home. My visit to the Marshall Trover treasure trove of classic saws was equally satisfying, if all too short. Dave Challenger allowed me to copy from his remaining collection of articles and magazines. Also helping along the way were Jim Kapp of Strata-G Communications, Gail Kenworthy of Stihl Canada, Ed Parkes and Terry Curtis of Husqvarna/Electrolux Canada, Therése Engström of Husqvarna Sweden, Herb of Deep Creek Tool Museum, Wayne Lemmond, Fred Whyte and Wayne Sutton of Stihl USA, Linda Challenger, Bill Orton, Roy Byers, Les Fowler, Gail Heffernan of Blount Canada, Petra Möck, Linda Bauer and Michael Hartmann of Andreas Stihl AG & Co. KG, Germany, Stan Fisher of Oregon Cutting Systems USA, Ian Radforth of the University of Toronto, Sam Catalano of Husqvarna North America, Nico Henkens, Greg MacDougall, Dianne Sullivan of Hatton-Brown Publishers Inc., Matt Hurst of Solo Inc., Yury A. Aleshin of the Russian Federation in Canada and Adrienne O'Mara of Archives New Zealand. Herb Lee and Jacquie Lee, in Vernon and Burnaby respectively, put me up for extended periods while I researched this book. Certainly last but not least my patient sons Malcolm and Simon and my wife Maureen, who has somehow stayed with me for over twenty years of "Hey, I just got this great idea…."
–David Lee

Publisher's Acknowledgements

More than normally, this book has been a collaborative effort with contributions from saw lovers that go beyond the ordinary. First and foremost, our two super-collectors Mike Acres and Marshall Trover opened their great collections to our photography team and spent days hefting and assembling stored saws for photography in ways that can simply never be repaid. To both of you we can only offer our profoundest thanks and hope the final product in some way makes you feel your efforts were not in vain. We absolutely could not have done it without you. Mike has also been an important contributor to the text, supplying crucial research materials, personal knowledge, the timeline, and several short sections to fill in missing parts. Both he and Marshall have also been most patient with the tedious process of fact-checking and proofing—we hope we finally got all those hyphens in the right places, and if we didn't it's sure not for any lack of effort by the two of you! We also owe a special thanks to Camp Six Logging Museum—PO Box 340, Tacoma, WA 98401-0340—where most of the saws from Marshall Trover's collection were photographed in appropriate surroundings. Personal thanks to curator Peter White and Camp Six staff Jenna White, and Gary Chapman for their enthusiastic help and support. Thanks also to Jim and June Jung of Seattle, who were gracious enough to let our photography team have free access to their property, and even supplied us with cold drinks and ice cream. We also owe a thank you to Dave Snowden of Standard Building Supplies in Burnaby BC, for letting us photograph saws in their lumberyard. Collectors Lowell Boyd and Art Patterson helped to prep and move saws to various photo shoots, and supplied some saws for the book. Duane Zollo drove from Camas WA to Renton just to deliver the NSU/Ural saw now seen on page 209. Chris Trover helped us lug over a hundred saws to the shoot site. Thanks also to old-timers Olaf Fedje, Don Corner and Frank White for answering questions nobody else could. Finally our thanks to Vici Johnstone for her tireless efforts in bringing all the scattered parts of this project together in this gorgeous book.

Text Copyright © 2006 David Lee
Photographs © 2006 Harbour Publishing

Paperback edition 2020

2 3 4 5 — 24 23 22

Harbour Publishing Co. Ltd.
P.O. Box 219, Madeira Park, BC, V0N 2H0
www.harbourpublishing.com

Front cover: Graphic illustration recreated by Pamela Cambiazo, IEL Super Pioneer 51 from collector Mike Acres, photo Vici Johnstone. / Back cover: Wolf electric from collector Marshall Trover, photo Brian Morris. / Front inside cover: Liner P51 with Villiers engine, c. 1945, from collector Marshall Trover, photo Brian Morris. / Back inside cover: Spear & Jackson Champion Junior with 125cc Villiers engine, built in 1946, from collector Mike Acres, photo Vici Johnstone. / Original graphics supplied by Mike Acres and Marshall Trover. / David Lee author photo by Maureen Cochrane.

Text and cover design by Roger Handling, Terra Firma Digital Arts.
Printed and bound in China.

Harbour Publishing acknowledges the support of the Canada Council for the Arts, which last year invested $153 million to bring the arts to Canadians throughout the country. / *Nous remercions le Conseil des arts du Canada de son soutien. L'an dernier, le Conseil a investi 153 millions de dollars pour mettre de l'art dans la vie des Canadiennes et des Canadiens de tout le pays.*

We also gratefully acknowledge financial support from the Government of Canada and from the Province of British Columbia through the BC Arts Council.

Library and Archives Canada Cataloguing in Publication

Title: Chainsaws : a history / David Lee ; produced in conjunction with Mike Acres and the Chain Saw Collectors Corner.

Names: Lee, David, 1952- author.

Description: Paperback reprint. Originally published in 2006 by Harbour Publishing. | Includes bibliographical references and index.

Identifiers: Canadiana 20190207108 | ISBN 9781550179118 (softcover)

Subjects: LCSH: Chain saws—History. | LCSH: Chain saws—Pictorial works.

Classification: LCC TJ1233 .L43 2020 | DDC 621.9/3409—dc23